MULTIPLE-CHOICE & FREE-RESPONSE QUESTIONS IN PREPARATION FOR THE AP CHEMISTRY EXAMINATION

(FIFTH EDITION)

Peter E. Demmin
Amherst Central High School, Retired
Amherst, New York

David W. Hostage
The Taft School
Watertown, Connecticut

Printed in the U.S.A.

PREFACE

The purpose of this book is to help students demonstrate their knowledge of introductory chemistry by successful performance on the multiple-choice and free-response sections of the Advanced Placement Examination in Chemistry. Mastery of the topics and addressing the questions presented in this book will provide much practice in "thinking" about chemistry. Such "thinking" is quite likely to improve performance on the examination and on work throughout the year.

THE BOOK

The book consists of an introductory section plus twelve chapters, each corresponding to a major topic in the Advanced Placement course description with a concise presentation of essential information - 300 multiple-choice questions, 36 free-response questions and chemical reactions where appropriate. These are followed by four practice examinations each with 75 multiple-choice and 8 free-response questions covering the full spectrum of topics. The list of *Equations and Constants* as provided with the AP Examination is included for use in the free-response section (II), only. No reference tables are provided for use in the multiple-choice section section (I) of the exam. Where needed, necessary constants and similar information is provided as part of the question. A Periodic Table of the Elements is available for all parts of the examination.

MULTIPLE-CHOICE QUESTIONS

Each multiple-choice question has one right (or best) answer – the key – and four wrong (or at least not very good) answers – the distractors. We advise students to make use of all these "wrong" answers as well as the right answers. An excellent study session – especially when three or four students work together – results when effort is made to contrive questions to which the wrong answers are now "correct". Thus, for many of the over 500 multiple-choice questions provided, there is now a way to think about perhaps 1500 or more questions.

FREE-RESPONSE QUESTIONS

Each free-response question can have many equally correct answers that earn full credit. Responses often include diagrams, sketches, reactions, graphs, structural formulas and other methods of expressing information. For quantitative problems, students are directed to show methods used and steps taken to produce the answer. In general, numbers should be labeled. Partial credit is awarded for such effort when it is correct or reasonable. For other free-response questions, students are advised to use accurate and relevant information and to provide explanations that are clearly presented and organized to respond directly to the question. It is generally not helpful to merely restate the question as part of the response. However, some students use this technique as a valuable organizer for the work that follows.

Occasionally, we will solve a quantitative problem in the review section of one of our twelve chapters. More often, we will outline a mental strategy for you to use to design a solution that works for you. If you need to see an actual solution, you will find several hundred in the STUDENT'S SOLUTIONS MANUAL covering all twelve topics.

USE OF CALCULATORS

Any calculations required for the multiple-choice questions are based on simple arithmetic and convenient algebra. Use of dimensional analysis – known also as "the factor-label method" or even "multiplying by ones" – simplifies the solution to many of these problems. Note that calculators are not permitted for the multiple-choice section of the examination. Students should NOT use a calculator for short, multiple-choice problems during the school year since that option will not be available during the examination.

Whatever quality is found in this work is due primarily to our observations of the learning and study skills used by our students over the years. We have tried to produce each question as a carefully designed exercise to help students learn – not merely to test the success of some previous learning experience. Thus, our greatest thanks goes to some five thousand youthful, vigorous minds. Sincere thanks also go to our colleagues, Patsy Mueller of Regina Dominican High School (Wilmette, IL) and Martin Fossett of Montclair (NJ) Kimberley Academy whose frank and careful review of the manuscript helped prevent some poor questions from getting any ink. Their insightful knowledge of chemistry and, especially, of how students perceive chemistry, provided significant input to the quality, level and clarity of the questions and their solutions. The experience of their years as active leaders and consultants in Advanced Placement Chemistry has made a major contribution to the credibility of this work offered for the preparation of students.

A large measure of gratitude goes to our spouses Ruth Demmin and Susan Hostage whose understanding and support helped carry us through many hours of writing a few hundred questions– only to find more of the same waiting their turn to be served.

Peter E. Demmin
Amherst Central High School, Retired
Amherst, New York

David W. Hostage
The Taft School
Watertown, Connecticut

All communications concerning this book should be addressed to the publisher and distributor:

D&S Marketing Systems, Inc.
1205 38th Street
Brooklyn, NY 11218

TABLE OF CONTENTS

PERIODIC TABLE OF THE ELEMENTS

1																	2
H 1.0079																	**He** 4.0026
3 **Li** 6.941	4 **Be** 9.012											5 **B** 10.811	6 **C** 12.011	7 **N** 14.007	8 **O** 16.00	9 **F** 19.00	10 **Ne** 20.179
11 **Na** 22.99	12 **Mg** 24.30											13 **Al** 26.98	14 **Si** 28.09	15 **P** 30.974	16 **S** 32.06	17 **Cl** 35.453	18 **Ar** 39.948
19 **K** 39.10	20 **Ca** 40.08	21 **Sc** 44.96	22 **Ti** 47.90	23 **V** 50.94	24 **Cr** 52.00	25 **Mn** 54.938	26 **Fe** 55.85	27 **Co** 58.93	28 **Ni** 58.69	29 **Cu** 63.55	30 **Zn** 65.39	31 **Ga** 69.72	32 **Ge** 72.59	33 **As** 74.92	34 **Se** 78.96	35 **Br** 79.90	36 **Kr** 83.80
37 **Rb** 85.47	38 **Sr** 87.62	39 **Y** 88.91	40 **Zr** 91.22	41 **Nb** 92.91	42 **Mo** 95.94	43 **Tc** (98)	44 **Ru** 101.1	45 **Rh** 102.91	46 **Pd** 106.42	47 **Ag** 107.87	48 **Cd** 112.41	49 **In** 114.82	50 **Sn** 118.71	51 **Sb** 121.75	52 **Te** 127.60	53 **I** 126.91	54 **Xe** 131.29
55 **Cs** 132.91	56 **Ba** 137.33	57 ***La** 138.91	72 **Hf** 178.49	73 **Ta** 180.95	74 **W** 183.85	75 **Re** 186.21	76 **Os** 190.2	77 **Ir** 192.2	78 **Pt** 195.08	79 **Au** 196.97	80 **Hg** 200.59	81 **Tl** 204.38	82 **Pb** 207.2	83 **Bi** 208.98	84 **Po** (209)	85 **At** (210)	86 **Rn** (222)
87 **Fr** (223)	88 **Ra** 226.02	89 **†Ac** 227.03	104 **Rf** (261)	105 **Db** (262)	106 **Sg** (263)	107 **Bh** (262)	108 **Hs** (265)	109 **Mt** (266)	110 § (269)	111 § (272)	112 § (277)						

§Not yet named

*Lanthanide Series

58 **Ce** 140.12	59 **Pr** 140.91	60 **Nd** 144.24	61 **Pm** (145)	62 **Sm** 150.4	63 **Eu** 151.97	64 **Gd** 157.25	65 **Tb** 158.93	66 **Dy** 162.50	67 **Ho** 164.93	68 **Er** 167.26	69 **Tm** 168.93	70 **Yb** 173.04	71 **Lu** 174.97

†Actinide Series

90 **Th** 232.04	91 **Pa** 231.04	92 **U** 238.03	93 **Np** 237.05	94 **Pu** (244)	95 **Am** (243)	96 **Cm** (247)	97 **Bk** (247)	98 **Cf** (251)	99 **Es** (252)	100 **Fm** (257)	101 **Md** (258)	102 **No** (259)	103 **Lr** (260)

ADVANCED PLACEMENT CHEMISTRY EQUATIONS AND CONSTANTS

ATOMIC STRUCTURE

$$E = h\nu \qquad c = \lambda\nu$$

$$\lambda = \frac{h}{m\upsilon} \qquad p = m\upsilon$$

$$E_n = \frac{-2.178 \times 10^{-18}}{n^2} \text{ joule}$$

EQUILIBRIUM

$$K_a = \frac{[H^+][A^-]}{[HA]}$$

$$K_b = \frac{[OH^-][HB^+]}{[B]}$$

$$K_w = [OH^-][H^+] = 1.0 \times 10^{-14} \text{ @ } 25°C$$

$$= K_a \times K_b$$

$$pH = -\log[H^+], \ pOH = -\log[OH^-]$$

$$14 = pH + pOH$$

$$pH = pK_a + \log\frac{[A^-]}{[HA]}$$

$$pOH = pK_b + \log\frac{[HB^+]}{[B]}$$

$$pK_a = -\log K_a, \ pK_b = -\log K_b$$

$$K_p = K_c(RT)^{\Delta n},$$

where Δn = moles product gas − moles reactant gas

THERMOCHEMISTRY/KINETICS

$$\Delta S° = \sum S° \text{ products} - \sum S° \text{ reactants}$$

$$\Delta H° = \sum \Delta H_f° \text{ products} - \sum \Delta H_f° \text{ reactants}$$

$$\Delta G° = \sum \Delta G_f° \text{ products} - \sum \Delta G_f° \text{ reactants}$$

$$\Delta G° = \Delta H° - T\Delta S°$$

$$= -RT \ln K = -2.303 \, RT \log K$$

$$= -n \mathscr{F} E°$$

$$\Delta G = \Delta G° + RT \ln Q = \Delta G° + 2.303 \, RT \log Q$$

$$q = mc\Delta T$$

$$C_p = \frac{\Delta H}{\Delta T}$$

$$\ln[A]_t - \ln[A]_0 = -kt$$

$$\frac{1}{[A]_t} - \frac{1}{[A]_0} = kt$$

$$\ln k = \frac{-E_a}{R}\left(\frac{1}{T}\right) + \ln A$$

E = energy $\qquad \upsilon$ = velocity

ν = frequency $\qquad n$ = principal quantum number

λ = wavelength $\qquad m$ = mass

p = momentum

Speed of light, $c = 3.0 \times 10^8 \text{ m s}^{-1}$

Planck's constant, $h = 6.63 \times 10^{-34} \text{ J s}$

Boltzmann's constant, $k = 1.38 \times 10^{-23} \text{ J K}^{-1}$

Avogadro's number $= 6.022 \times 10^{23} \text{ mol}^{-1}$

Electron charge, $e = -1.602 \times 10^{-19}$ coulomb

1 electron volt per atom $= 96.5 \text{ kJ mol}^{-1}$

Equilibrium Constants

K_a (weak acid)

K_b (weak base)

K_w (water)

K_p (gas pressure)

K_c (molar concentrations)

$S°$ = standard entropy

$H°$ = standard enthalpy

$G°$ = standard free energy

$E°$ = standard reduction potential

T = temperature

n = moles

m = mass

q = heat

c = specific heat capacity

C_p = molar heat capacity at constant pressure

E_a = activation energy

k = rate constant

A = frequency factor

Faraday's constant, $\mathscr{F} = 96,500$ coulombs per mole of electrons

Gas constant, $R = 8.31 \text{ J mol}^{-1} \text{ K}^{-1}$

$$= 0.0821 \text{ L atm mol}^{-1} \text{ K}^{-1}$$

$$= 8.31 \text{ volt coulomb mol}^{-1} \text{ K}^{-1}$$

GASES, LIQUIDS, AND SOLUTIONS

$$PV = nRT$$

$$\left(P + \frac{n^2 a}{V^2}\right)(V - nb) = nRT$$

$$P_A = P_{total} \times X_A, \text{ where } X_A = \frac{\text{moles A}}{\text{total moles}}$$

$$P_{total} = P_A + P_B + P_C + \ldots$$

$$n = \frac{m}{M}$$

$$\text{K} = {}^{\circ}\text{C} + 273$$

$$\frac{P_1 V_1}{T_1} = \frac{P_2 V_2}{T_2}$$

$$D = \frac{m}{V}$$

$$u_{rms} = \sqrt{\frac{3kT}{m}} = \sqrt{\frac{3RT}{M}}$$

$$KE \text{ per molecule} = \frac{1}{2} mv^2$$

$$KE \text{ per mole} = \frac{3}{2} RT$$

$$\frac{r_1}{r_2} = \sqrt{\frac{M_2}{M_1}}$$

molarity, M = moles solute per liter solution

molality = moles solute per kilogram solvent

$$\Delta T_f = iK_f \times \text{molality}$$

$$\Delta T_b = iK_b \times \text{molality}$$

$$\pi = MRT$$

$$A = abc$$

P = pressure
V = volume
T = temperature
n = number of moles
D = density
m = mass
v = velocity

u_{rms} = root-mean-square speed
KE = kinetic energy
r = rate of effusion
M = molar mass
π = osmotic pressure
i = van't Hoff factor
K_f = molal freezing-point depression constant
K_b = molal boiling-point elevation constant
A = absorbance
a = molar absorptivity
b = path length
c = concentration
Q = reaction quotient
I = current (amperes)
q = charge (coulombs)
t = time (seconds)
E° = standard reduction potential
K = equilibrium constant

OXIDATION-REDUCTION; ELECTROCHEMISTRY

$$Q = \frac{[\text{C}]^c [\text{D}]^d}{[\text{A}]^a [\text{B}]^b}, \text{ where } a\,\text{A} + b\,\text{B} \rightarrow c\,\text{C} + d\,\text{D}$$

$$I = \frac{q}{t}$$

$$E_{cell} = E^{\circ}_{cell} - \frac{RT}{n\mathscr{F}} \ln Q = E^{\circ}_{cell} - \frac{0.0592}{n} \log Q \text{ @ 25°C}$$

$$\log K = \frac{nE^{\circ}}{0.0592}$$

Gas constant, R = 8.31 J mol^{-1} K^{-1}

= 0.0821 L atm mol^{-1} K^{-1}

= 8.31 volt coulomb mol^{-1} K^{-1}

Boltzmann's constant, k = 1.38×10^{-23} J K^{-1}

K_f for H_2O = 1.86 K kg mol^{-1}

K_b for H_2O = 0.512 K kg mol^{-1}

1 atm = 760 mm Hg

= 760 torr

STP = 0.000°C and 1.000 atm

Faraday's constant, $\mathscr{F}$ = 96,500 coulombs per mole of electrons

STANDARD REDUCTION POTENTIALS IN AQUEOUS SOLUTION AT 25°C

Half-reaction			$E°(V)$
$F_2(g) + 2\,e^-$	$\rightarrow$	$2\,F^-$	2.87
$Co^{3+} + e^-$	$\rightarrow$	Co^{2+}	1.82
$Au^{3+} + 3\,e^-$	$\rightarrow$	$Au(s)$	1.50
$Cl_2(g) + 2\,e^-$	$\rightarrow$	$2\,Cl^-$	1.36
$O_2(g) + 4\,H^+ + 4\,e^-$	$\rightarrow$	$2\,H_2O(l)$	1.23
$Br_2(l) + 2\,e^-$	$\rightarrow$	$2\,Br^-$	1.07
$2\,Hg^{2+} + 2\,e^-$	$\rightarrow$	Hg_2^{2+}	0.92
$Hg^{2+} + 2\,e^-$	$\rightarrow$	$Hg(l)$	0.85
$Ag^+ + e^-$	$\rightarrow$	$Ag(s)$	0.80
$Hg_2^{2+} + 2\,e^-$	$\rightarrow$	$2\,Hg(l)$	0.79
$Fe^{3+} + e^-$	$\rightarrow$	Fe^{2+}	0.77
$I_2(s) + 2\,e^-$	$\rightarrow$	$2\,I^-$	0.53
$Cu^+ + e^-$	$\rightarrow$	$Cu(s)$	0.52
$Cu^{2+} + 2\,e^-$	$\rightarrow$	$Cu(s)$	0.34
$Cu^{2+} + e^-$	$\rightarrow$	Cu^+	0.15
$Sn^{4+} + 2\,e^-$	$\rightarrow$	Sn^{2+}	0.15
$S(s) + 2\,H^+ + 2\,e^-$	$\rightarrow$	$H_2S(g)$	0.14
$2\,H^+ + 2\,e^-$	$\rightarrow$	$H_2(g)$	0.00
$Pb^{2+} + 2\,e^-$	$\rightarrow$	$Pb(s)$	−0.13
$Sn^{2+} + 2\,e^-$	$\rightarrow$	$Sn(s)$	−0.14
$Ni^{2+} + 2\,e^-$	$\rightarrow$	$Ni(s)$	−0.25
$Co^{2+} + 2\,e^-$	$\rightarrow$	$Co(s)$	−0.28
$Tl^+ + e^-$	$\rightarrow$	$Tl(s)$	−0.34
$Cd^{2+} + 2\,e^-$	$\rightarrow$	$Cd(s)$	−0.40
$Cr^{3+} + e^-$	$\rightarrow$	Cr^{2+}	−0.41
$Fe^{2+} + 2\,e^-$	$\rightarrow$	$Fe(s)$	−0.44
$Cr^{3+} + 3\,e^-$	$\rightarrow$	$Cr(s)$	−0.74
$Zn^{2+} + 2\,e^-$	$\rightarrow$	$Zn(s)$	−0.76
$Mn^{2+} + 2\,e^-$	$\rightarrow$	$Mn(s)$	−1.18
$Al^{3+} + 3\,e^-$	$\rightarrow$	$Al(s)$	−1.66
$Be^{2+} + 2\,e^-$	$\rightarrow$	$Be(s)$	−1.70
$Mg^{2+} + 2\,e^-$	$\rightarrow$	$Mg(s)$	−2.37
$Na^+ + e^-$	$\rightarrow$	$Na(s)$	−2.71
$Ca^{2+} + 2\,e^-$	$\rightarrow$	$Ca(s)$	−2.87
$Sr^{2+} + 2\,e^-$	$\rightarrow$	$Sr(s)$	−2.89
$Ba^{2+} + 2\,e^-$	$\rightarrow$	$Ba(s)$	−2.90
$Rb^+ + e^-$	$\rightarrow$	$Rb(s)$	−2.92
$K^+ + e^-$	$\rightarrow$	$K(s)$	−2.92
$Cs^+ + e^-$	$\rightarrow$	$Cs(s)$	−2.92
$Li^+ + e^-$	$\rightarrow$	$Li(s)$	−3.05

BOOSTING YOUR EXAM SCORE – GETTING STARTED

The purpose of this book is to help you earn a high score on the AP CHEMISTRY EXAMINATION. We assume that you have received adequate instruction throughout the course and that you have a textbook in which you can find information and problem-solving methods.

We are focused on the Examination. We will help you

- choose among options for Multiple-Choice Questions

- design answers to Free-Response Questions

In each of our twelve chapters, we have identified what you need to know to produce a great score on the Examination. In these chapters, we will provide ample opportunity for you to practice choosing and designing answers. We will not provide a book with questions and answers that offer the temptation to simply read and agree that, once again, that author has surely presented a great answer. We work from the premise that you must produce the answer. This book gives you about 1000 AP-exam-style chances to produce answers by choosing and designing yourself!

Most often, we will tell what to think about and what to do, rather than showing you how to do it. You can find useful diagrams, charts and illustrations of how-to-do-it problem-solving in your textbook. Use our book to direct the work you must do for yourself. Specifically, we will help you

- choose the best answer to each of the multiple choice questions – and also figure out why the incorrect answers are wrong. Chances are that a "wrong" answer will be the right answer to some future question.

- write answers to the free response questions that adequately represent what you know

- write chemical reactions that illustrate your knowledge of how substances behave

- solve quantitative problems using a persuasive method that can be understood by someone else.

This is not a book full of questions and answers to be read by the student and somehow *learned*. It is a book full of opportunity to become the master of what you already know about chemistry.

DEALING WITH THE EXAMINATION

Section I Multiple Choice Questions – choosing the best answer

Don't waste time. Read through each question as rapidly as possible. If necessary, avoid the questions that look lengthy – that have too many words. Try to formulate a good answer before reading the five choices. When your good answer matches one of the choices, chances are you have a winner. If you have confidence in that "winner", "blacken the box" on the answer sheet and move on. If you have some uncertainty about your response, circle your best choice on the question paper. Get through as much of the exam as possible, then return for another look at your uncertain responses.

On the other hand, if you're like "I have no clue!", skip that question. "Guess" only if you can exclude two or even three of the choices with some degree of confidence.

Should you guess?

Try this imaginary question:

What is the symbol for tungsten?

(A) Te   (B) K   (C) Os   (D) W   (E) Fe

Chances are that you have used K and Fe often in your AP Chemistry career. You know neither potassium (K) nor iron (Fe) is tungsten.

- It might be Te. ("Tungsten" starts with "T" and has "e".)

- It might be Os.

   (Perhaps the Romans found tungsten in the shells of osprey eggs.)

- It might be W.

   (Perhaps *walium* is to tungsten as *kalium* is to potassium.)

By ruling out K and Fe, guessing makes sense statistically. You have 1/3 chance to get it right. If you're wrong, you only lose 1/4 of a point. Actually, W stands for wolfram, an old German name for tungsten, obtained from the ore mineral, wolframite – but you get the point.

In Section I on recent exams, a net score of Right – 1/4 Wrong of about 45 has produced a test score of 4 or 5 for 99% of the test-takers. Suppose you skipped 10 questions and answered 65. Getting 49 right (and 16 wrong) gives you a score of 45 and probably a test score of at least 4. If your sights are a little lower, to get a 3, you need only about 25 points on Section I.

Section II   Free response – designing a great answer

Use a good pen that doesn't bleed into porous paper. Avoid any pen that leaves a blop of ink when you change direction at the top of the letter ℓ. Avoid smeary pencils. Cross out anything you don't want to include in your answer. No grader (We call them readers.) will spend the time to read anything unnecessary. Believe it or not, the reader wants to give you points. That's what Free Response is all about.

Quantitative Problems

Use a well-organized problem-solving method. Effective application of dimensional analysis to any quantitative problem sends a very positive message to the reader. A jumble of unlabeled numbers cries out for no partial credit. Be sure to include the proper label on the answer. Draw a box around your final numerical answer. Watch for proper use of significant figures. Most problems are written such that two or three significant figures are expected in the answer. The number of significant figures in your answer should match the number provided in the given information.

The essay questions

You generally do not need to write anything resembling a traditional essay to get full credit. A few sentences, a diagram, a chemical equation or two, a list of pertinent facts or a brief explanation is usually satisfactory.

Avoid trying to restate the question as an attempt to earn points. This is easily recognized by the reader who is trained to look for an answer, not another version of the question.

The answers we provide in the STUDENT'S SOLUTIONS MANUAL are exceptionally thorough. They should not be regarded as the standard for full credit. To earn full credit, you can usually say a whole lot less than we do. Be sure to get your teacher or other students to read some of your "essays". That will help you learn how to write something that makes sense to others.

The Reactions

We're going to address the toughest part of Section II first – right now as we are getting started. Even though it is Question 4, it gets top priority in our efforts to boost your exam score. There is no need for you to ever leave anything blank in the five boxes you have available for question 4. Of the 15 points available, you don't need to settle for anything less than 8. If you adopt our system, you have a great shot at 12 and maybe even all 15 points. Chances are you already know a lot about how substances behave. We will show you how to organize that information for easy efficient access.

As you work through our twelve topic-specific chapters, you can use the skills you develop by writing chemical reactions right from the start. In every chapter, there is a chance to use information about how substances behave.

Get a set of sharp pencils with good erasers. Start writing reactions beginning on page 9. For some students, a study group of three or four is a great way to begin to organize your knowledge about chemical reactions. If your group can find some chalk and a blackboard (or more likely a whiteboard and some markers), you will make remarkable progress.

Some teachers and students think of this question as writing equations. A better description is the writing of chemical formulas and the prediction of products of chemical reactions.

This kind of prediction is NOT a random recall operation - a situation where you must try to remember some obscure fact, such as the date of the Battle of Hastings (1066) or the modern name of the tract of land known as Seward's Folly (Alaska). Predicting products can be successfully accomplished by adopting a mental system for thinking about chemical reactions. Such prediction is partly a test of how well you can classify or categorize the storage of your knowledge and then successfully draw upon that storehouse of knowledge.

Question 4, stated below, presents one of the greatest but most predictable challenges of the AP Chemistry Examination.

The Directions for Question 4

Write the formulas to show the reactants and the products for any FIVE of the laboratory situations described below. Answers to more than five choices will not be graded. In all cases a reaction occurs. Assume that solutions are aqueous unless otherwise indicated. Represent substances in solutions as ions if the substances are extensively ionized. Omit formulas for any ions or molecules that are unchanged by the reaction. You need not balance the equations.

Put another way, this means tell us what you know about the chemical properties of several elements, compounds or mixtures. Each "laboratory situation" is generally presented as a mixture of substances that undergo chemical change. Use the "laboratory situations" below to practice writing reactions by identifying the appropriate reaction category then predicting the products. The identification and descriptions of correct responses are found in the Students Solutions Manual beginning on page 208.

The CATEGORIES OF COMMONLY-USED REACTIONS – a brief list

Use this as a mental file system for storing knowledge of chemical properties. If you like a hard copy too, consider preparing nine 3x5 cards for study purposes.

REDOX

1. Synthesis (Direct Combination)
2. Decomposition (Analysis)
3. Single Replacement
4. Burning (Simple combustion)
5. Redox by commonly-used "agents"

No REDOX

6. Double Replacement
7. Acid-Base Reactions
8. Complexation

Organic Reactions

9. Reactions that illustrate chemical properties of organic substances

CATEGORIES OF COMMONLY-USED REACTIONS – that same list: expanded

Consider making nine 4x6 cards for study purposes.

REDOX

1. Synthesis (Direct Combination of elements)

   oxidizing agent – usually nonmetals (halogens, C, S, N, O, P)

   reducing agent – metal or less electronegative nonmetal

2. Decomposition (Analysis)

   heating, electrolysis

3. Single Replacement

   refer to an activity series, a table of E° values or even groups of elements from the Periodic Table

4. Burning (Simple combustion)

   metal, non-metal ignited in oxygen or air to form oxides of all elements present; hydrocarbon or other organic compound ignited in oxygen or air to form H_2O and CO_2 (or CO or C)

5. Redox by commonly-used "agents"

   oxidizing agents: MnO_4^- (acid solution $\rightarrow Mn^{2+}$)

                               (basic solution $\rightarrow MnO_2$ or MnO_4^{2-})

                        $Cr_2O_7^{2-}$ (acid solution $\rightarrow Cr^{3+}$)

                        H_2O_2 ($\rightarrow H_2O$)

                        NO_3^- (acid solution $\rightarrow NO_2$, NO, NH_4^+)

                        I_2 or other nonmetal ($\rightarrow I^-$ or other anion)

   reducing agents: Ca or other metal ($\rightarrow Ca^{2+}$ or other cation)

                        SO_2 (acid solution $\rightarrow SO_4^{2-}$)

                        Sn^{2+} ($\rightarrow Sn^{4+}$)

No REDOX

6. Double Replacement

   a. precipitation – see solubility rules; includes reactions often seen in qualitative analysis, especially Ag^+, Hg_2^{2+}, Pb^{2+}

   b. formation of an insoluble gas, usually CO_2 or SO_2

   c. formation of water from sources of H^+ and OH^- – "Arrhenius neutralization"

7. Acid-Base Reactions

    a. Arrhenius – neutralization of acids and bases as above

    b. Bronsted-Lowry – proton transfer; conjugate acid/base pairs; hydrolysis of cations and anions in salts; amphiprotic behavior

    c. Lewis – donating/accepting share in a pair of electrons

    d. amphiprotic species – donating/accepting protons by same species such as the ions HCO_3^- and HSO_4^- and the precipitates $Al(OH)_3$ and $Zn(OH)_2$

    e. anhydrides – oxides of metals and nonmetals dissolved in water to form acids and bases

8. Complexation – a specific illustration of Lewis acid-base behavior

    common ligands (Lewis bases): NH_3, $C_2O_4^{2-}$, CN^-, H_2O, OH^-, Cl^-

    Lewis acids: $AlCl_3$, BF_3, ions of transition elements

Organic Reactions

9. Chemical reactions of organic compounds

    a. addition: unsaturated hydrocarbon plus a halogen or hydrogen halide forms haloalkane or other saturated hydrocarbon

    b. substitution: saturated hydrocarbon plus a halogen or hydrogen halide forms haloalkane

    c. esterification: organic acid plus alcohol yields ester and water

    d. polymerization: monomers linked to form a polymer

    e. oxidation: alcohol oxidized to ketone, aldehyde or acid

---

SOLUBILITY RULES for Category 6a

1. All common salts of the Group 1 (alkali metals) elements and the NH_4^+ ion are soluble.

2. All common nitrates and acetates are soluble.

3. All binary compounds of the Group 17 (halogen) elements (other than F) with metals are soluble except those of silver, mercury(I) and lead.

4. All sulfates are soluble except those of barium, strontium, calcium, silver, lead and mercury.

5. Except for those in Rule 1, carbonates, hydroxides, oxides and phosphates are insoluble.

The Process of Product Prediction (See also our flowchart on page 8.)

A.   Does the mixture of reactants include an organic compound?

     YES = choose a component of category 9 and "predict" the product(s)

     NO = go to part B

B.   Does the mixture of reactants suggest oxidation or reduction? Is there likely to be a change in oxidation number?

     YES = choose a category below and predict the products(s)

        1.   direct combination (synthesis) two elements → one compound

        2.   decomposition (analysis): heating, electrolyzing or other addition of energy to one compound constituent elements or simpler compound(s)

        3.   single replacement: element plus compound → different element and different compound

        4.   burning (combustion): element or compound ignited in air, oxygen or other oxidizing environment usually in the gas phase → an oxide

        5.   redox reaction: mixture includes a commonly used oxidizing agent or reducing agent

     NO = go to part C

C.   Is this a mixture with all components in water solution?

     YES = choose a category below

        6.   double replacement – ions trade "partners" with precipitation or gas formation

        7.   complexation – a coordination compound is formed

        8.   acid-base reactions (Note: Some acid base reactions do not require a water solution environment.) - neutralization or proton transfer

     NO = you have probably missed a key component of the reaction mixture; start again at part A or choose another reaction mixture from Question 4.

The Process of Product Prediction

For nearly any reaction mixture presented in Question 4

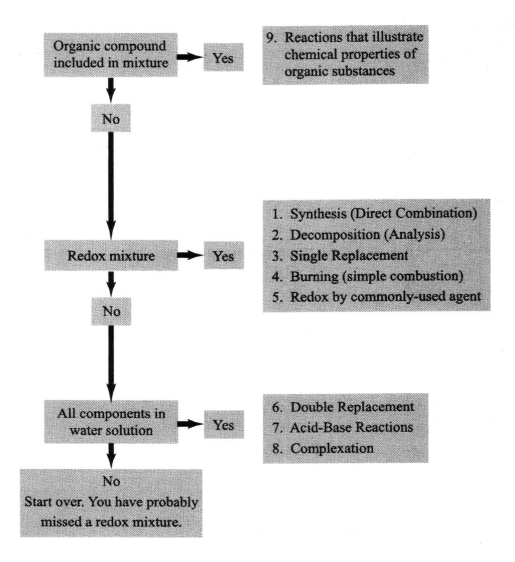

Note that we have not given you a very scary and lengthy list of reactions to remember. You will make your own list on the next five pages.

To get started we suggest you use a pencil with a good eraser. Change to a pen once you become a master of this process.

Practice Writing Twenty Chemical Reactions

1. A dilute solution of potassium dichromate is added to a solution of methanoic (formic) acid.

    The Process: (A) YES NO   (B) YES NO   (C) YES NO

    The category – number and name: _____ _____

    The reaction:

2. A solution of potassium iodide is added to an acidified solution of potassium iodate.

    The Process: (A) YES NO   (B) YES NO   (C) YES NO

    The category – number and name: _____ _____

    The reaction:

3. A few crystals of sodium permanganate are added to an acidified solution of sodium sulfite.

    The Process: (A) YES NO   (B) YES NO   (C) YES NO

    The category – number and name: _____ _____

    The reaction:

4. A few crystals of lithium hydride are added to water.

    The Process: (A) YES NO   (B) YES NO   (C) YES NO

    The category – number and name: _____ _____

    The reaction:

5. A solution of iron(II) sulfate is added to a solution of acidified potassium dichromate.

    The Process: (A) YES NO   (B) YES NO   (C) YES NO

    The category – number and name: _____ _____

    The reaction:

6. Equal volumes of equimolar solutions of hydrochloric acid and sodium phosphate are mixed.

   The Process: (A) YES NO   (B) YES NO   (C) YES NO

   The category – number and name: _____ _____

   The reaction: [   ]

7. Equal volumes of equimolar solutions of hydrochloric acid and sodium hydrogen phosphate are mixed.

   The Process: (A) YES NO   (B) YES NO   (C) YES NO

   The category – number and name: _____ _____

   The reaction: [   ]

8. Equal volumes of equimolar solutions of hydrochloric acid and sodium dihydrogen phosphate are mixed.

   The Process: (A) YES NO   (B) YES NO   (C) YES NO

   The category – number and name: _____ _____

   The reaction: [   ]

9. A few crystals of sodium fluoride are added to a dilute solution of hydrochloric acid.

   The Process: (A) YES NO   (B) YES NO   (C) YES NO

   The category – number and name: _____ _____

   The reaction: [   ]

10. A few crystals of potassium carbonate are added to a dilute solution of hydrochloric acid.

    The Process: (A) YES NO   (B) YES NO   (C) YES NO

    The category – number and name: _____ _____

    The reaction: [   ]

11. A few drops of a solution of mercury(I) nitrate are added to a solution of sodium carbonate.

    The Process: (A) YES NO   (B) YES NO   (C) YES NO

    The category – number and name: _____ _____

    The reaction: [   ]

12. A solution of barium hydroxide is added to a solution of ammonium sulfate and warmed.

       The Process: (A) YES NO   (B) YES NO   (C) YES NO

       The category – number and name: _____ _____

       The reaction:

13. A few crystals of calcium fluoride are added to hot concentrated sulfuric acid.

       The Process: (A) YES NO   (B) YES NO   (C) YES NO

       The category – number and name: _____ _____

       The reaction:

14. Chlorine gas is bubbled into a solution of potassium iodide.

       The Process: (A) YES NO   (B) YES NO   (C) YES NO

       The category – number and name: _____ _____

       The reaction:

15. A solution of sodium hydroxide is mixed with a solution of ethanoic (acidic) acid.

       The Process: (A) YES NO   (B) YES NO   (C) YES NO

       The category – number and name: _____ _____

       The reaction:

16. A solution of potassium permanganate is added to an acidified solution of hydrogen peroxide.

       The Process: (A) YES NO   (B) YES NO   (C) YES NO

       The category – number and name: _____ _____

       The reaction:

17. A mixture of methane and excess chlorine gas is irradiated with ultraviolet light.

       The Process: (A) YES NO   (B) YES NO   (C) YES NO

       The category – number and name: _____ _____

       The reaction:

18. Ethanol is mixed with methanoic (formic) acid in the presence of concentrated sulfuric acid.

        The Process: (A) YES NO   (B) YES NO   (C) YES NO

        The category – number and name: _____ _____

        The reaction:

19. Ethene is heated in an atmosphere of bromine vapor.

        The Process: (A) YES NO   (B) YES NO   (C) YES NO

        The category – number and name: _____ _____

        The reaction:

20. Excess concentrated ammonia solution is added to a suspension of zinc hydroxide.

        The Process: (A) YES NO   (B) YES NO   (C) YES NO

        The category – number and name: _____ _____

        The reaction:

21. Excess concentrated hydrochloric acid is added to 3 M aluminum nitrate.

        The Process: (A) YES NO   (B) YES NO   (C) YES NO

        The category – number and name: _____ _____

        The reaction:

22. Solid ammonium carbonate is heated in an open vessel.

        The Process: (A) YES NO   (B) YES NO   (C) YES NO

        The category – number and name: _____ _____

        The reaction:

23. A few crystals of strontium oxide are added to water.

        The Process: (A) YES NO   (B) YES NO   (C) YES NO

        The category – number and name: _____ _____

        The reaction:

24. A small quantity of manganese(IV) oxide is mixed with solid potassium chlorate and heated.

The Process: (A) YES NO   (B) YES NO   (C) YES NO

The category – number and name: _____ _____

The reaction:

25. Sulfur dioxide gas is bubbled into water.

The Process: (A) YES NO   (B) YES NO   (C) YES NO

The category – number and name: _____ _____

The reaction:

CHAPTER 1
ATOMIC STRUCTURE

ATOMS AND THE ATOMIC THEORY

According to modern atomic theory, the chemical elements exist as fundamental, indivisible particles with properties that are identical for every atom of each element and different from the properties of every other element. While the discovery of subatomic particles, wave phenomena, isotopes, and other observations makes this claim, strictly speaking, false, it is close enough to the "truth" to be useful in the sciences. You should be prepared to discuss the history and evolution of the atomic theory with attention to the experiments that led to important advances. Dates are not important; understanding the experiments and their related conclusions is.

Scientist	Significant Work
Dalton	Law of Multiple Proportions → individual identical particles
Crookes	cathode rays → atoms have positive and negative components
Thomson	beam of electrons → charge-to-mass ratio for electrons
Millikan	oil drop experiment → each electron has the same discrete charge
Rutherford	alpha bombardment of atoms → atoms are mostly empty space with dense nuclei
DeBroglie & Planck	matter has wave-like properties → energy changes in atoms are quantized
Bohr	hydrogen emission spectra → planetary model of the atom
Many workers	combining efforts → charge cloud model

What's a Theory
Hypothesis/Law/Theory/Experiment

If you discuss hypotheses, laws and theories in your exam essays, be sure to use the science-related perspective. Because these words are in everyday use, their connotations or even definitions have strayed from scientific meaning.

Observations, especially when obtained using scientifically designed **experiments** based on carefully specified hypotheses, are sometimes restated as **laws** summarizing these observations. Many coordinated observations, **hypotheses**, and experiments generate **theories**. A theory, such as the Atomic Theory, explains or accounts for a broad set of behaviors of matter or living organisms in the real world. Another theory is the Germ Theory of Disease.

Theories are not guesses based on whimsy, bias or opinion. When a theory has stood the test of time and many experiments, it becomes widely accepted in the scientific community. It is common to encounter competing theories, especially in emerging fields of study.

In the world of science, the closest thing to a guess is an **hypothesis**. One proposes hypotheses to be tested experimentally. Many tested hypotheses become known collectively as a theory. Observations and laws tell what happens in the real world. Experiments are guides to asking the right questions about what happens. After we have watched what happens long enough, we acquire sufficient confidence to offer a theory; that is, we understand a phenomenon well enough to explain why something happened, confident enough to make a prediction that it will happen again. Note that each prediction presents another test of the theory.

The Charge Cloud Model

Current atomic theory is based upon the charge cloud model in which a positive nucleus is surrounded by electrons. The **protons** and **neutrons** found in the nucleus (**nucleons**) account for **atomic number** and **mass number** as well as the charge on the nucleus. They also account for **atomic mass**, including **weighted average atomic mass**, based on distribution of naturally-occurring isotopes. It is the differences in number of neutrons that accounts for the existence of **isotopes** for nearly every element. The electrons surround the nucleus as particles – or perhaps waves (the wave/particle duality) – and are assigned to **orbitals**, **sublevels** and **energy levels**. It is the movement of electrons between energy levels and sublevels that accounts for atomic spectra – both emission and absorption spectra. You must be able to connect the orbital (spectrographic) notation for electrons to the corresponding **quantum numbers** as shown in Figure 1.1. You must also be able to sketch the electron distribution associated with s and p orbitals. In general, d and f orbitals are not easily represented in two dimensional diagrams.

Figure 1.1 Ranges of Values of n, ℓ, and m, through $n = 4$

n	Possible Values of ℓ	Subshell Designation	Possible Values of m (or m_ℓ)	Number of Orbitals in Subshell	Total Number of Orbitals in Shell
1	0	$1s$	0	1	1
2	0	$2s$	0	1	
	1	$2p$	1, 0, –1	3	4
3	0	$3s$	0	1	
	1	$3p$	1, 0, –1	3	
	2	$3d$	2, 1, 0, –1, –2	5	9
4	0	$4s$	0	1	
	1	$4p$	1, 0, –1	3	
	2	$4d$	2, 1, 0, –1, –2	5	
	3	$4f$	3, 2, 1, 0, –1, –2, –3	7	16

THE PERIODIC TABLE OF THE ELEMENTS

It is the structure of an atom, particularly the arrangement of its electrons in the extra-nuclear (that means *outside of the nucleus*) region, which determines where an element fits on the Periodic Table. A Periodic Table (See Page i.) is provided for use on all parts of the AP exam. Each of the seven horizontal lines on the Periodic Table contains a "period" of elements. Each period of elements begins with an Alkali Metal and ends with a Noble Gas. There are several families found as vertical columns on the Periodic Table. Members of the same family have similar chemical and physical properties because they have the same electron configuration in the outer energy level. Such families include the Alkali Metals, Alkaline Earth Metals, Pnictides, Chalcogens, Halogens and Noble Gases. Not every element is assigned to a family. Another category of elements is the transition elements, found at the trough of the table. Note that the Lanthanides and the Actinides are identified on the Periodic Table provided with the AP Examination.

Properties of the elements and the Periodic Table

The Periodic Table can be used to represent similarities and differences in physical and chemical properties of the elements. You need to know the definition of each property listed below. You must also be able to discuss the trends in each property in terms of location on the Periodic Table. See Figure 1.2.

Properties of the elements as atoms and ions

Atomic radius   distance from the nucleus to the "edge" of an atom, expressed in nm

Ionization energy   energy required to remove the most loosely held electron from a neutral atom to form an ion in the gas phase with a charge of 1+

$$X_{(g)} \rightarrow X^+{}_{(g)} + e^-$$

Electron affinity   energy given off from an atom when an electron enters the available (vacant) orbital with the lowest energy to form an ion in the gas phase with charge of 1-

$$X_{(g)} + e^- \rightarrow X^-{}_{(g)}$$

Metallic character   the extent to which an element acts like a metal

- **Physical properties**   high electrical and thermal conductivity; malleable, ductile and sectile (pound, stretch, cut) but not brittle; lustrous

- **Chemical properties**   metallic bonding as elements; form positive ions when reacting with non-metals to form compounds

Electronegativity   the relative attractive force for a shared pair of electrons; Pauling scale runs from 4.0 to 0.7

Ionic radius   distance from the center of the nucleus of an ion to its "edge"; expressed in nanometers

Figure 1.2 Trends in properties: periodic relationships

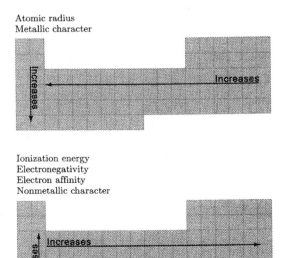

Ionic radius within a group

In general the size of any ion is determined by the number of protons in the nucleus which attract the electrons surrounding the nucleus. The trend of ionic radius within a period is not easily summarized. The trend in ionic radius for isoelectronic (that means equal, *iso*, electronic structure) species is such that greater ionic radius is associated with lesser nuclear charge. See Figure 1.3 for the radii of ions with 20 electrons.

Figure 1.3 Ionic Radii for ions with 20 electrons

Ion	$_{15}P^{3-}$	$_{16}S^{2-}$	$_{17}Cl^{-}$	$_{19}K^{+}$	$_{20}Ca^{2+}$	$_{21}Sc^{3+}$
Radius, nm (nanometers)	0.212	0.184	0.181	0.133	0.099	0.073

Using the List of *Equations and Constants*

See page ii for the full List of *Equations and Constants*. This list is provided for use in Part II of the AP Examination Free Response. The first section contains identification of each term found in the section with equations for Atomic Structure. An annotated section from that list is given with each chapter where appropriate. A quantitative problem containing some of the terms from these equations with their numerical values may appear on the exam. Such a problem is likely to include all but one of the values for any equation. That value can be determined algebraically. The units specified for the given values will help you identify which equation is involved if you know the units for the terms in the equations which appear on the list below. The values for c and h are given elsewhere in this list as

Speed of light, $c = 3.00 \times 10^8$ m sec^{-1}

Planck's constant, $h = 6.63 \times 10^{-34}$ J sec

You need to know or be able to figure out the labels for each term in the List. Figuring out the labels is best. That way you avoid the phenomenon of *disremembering* - a word we use to mean remembering incorrectly with confidence.

from the TOPIC OUTLINE (website: apcentral.collegeboard.com)

I. Structure of Matter

A. Atomic theory and atomic structure

1. Evidence for the atomic theory

2. Atomic masses; determination by chemical and physical means

3. Atomic number and mass number, isotopes

4. Electron energy levels: atomic spectra, quantum numbers, atomic orbitals

5. Periodic relationships including, for example, atomic radii, ionization energies, electron affinities, oxidation states

from the list of CHEMICAL CALCULATIONS

No specified calculations are associated with Atomic Structure. However, the equations listed below appear on the *List* ... and are likely to be used in some calculations.

from the list of EQUATIONS & CONSTANTS

$E = h\nu$	energy in joules = (Planck's constant in joule · sec) × frequency in 1/sec or sec^{-1}
$c = \lambda\nu$	speed of light in m/sec = wave length in m (or nm) × frequency in sec^{-1}
$\lambda = h/mv$	wave length in m (or nm) = Planck's constant in joule sec ÷ (mass in kg × velocity in m sec^{-1}) (Recall that joule = kg m sec^{-1} m sec^{-1})
$p = mv$	momentum = kg m sec^{-1}
$E_n = \dfrac{-2.1789 \times 10^{-18}}{n^2}$ joule	energy of the nth energy level in joules

from the list of RECOMMENDED EXPERIMENTS

#17 Using the spectrophotometer

Because atoms (and ions) of elements have different absorption spectra, a spectrophotometer can be used to confirm their presence in a solution. Both qualitative and quantitative information can be obtained; that means, substance present? YES or NO, and, if present, at what concentration.

It is the structure of atoms that ultimately determines chemical behavior. You need to be alert to similarities and differences in order to account for a wide variety of experimental observations, results and conclusions.

Multiple Choice Questions

Questions 1-5: The set of lettered choices below is a list of regions of the electromagnetic spectrum. Select the lettered choice that best fits each statement. A choice may be used once, more than once, or not at all.

(A) infrared
(B) microwaves
(C) ultraviolet
(D) visible
(E) x-rays

1. Radiation in this region has the highest energy of those listed.

2. Radiation in this region is used to analyze colored solutions.

3. Radiation in this region passes through ordinary glass but can be blocked by treated glass.

4. Radiation in this region is used in the analysis of metallic crystal structures.

5. Radiation in this region is used as a detection beam across doorways and windows.

---

6. Which is a list of elements in order of **increasing** first ionization energy?
    (A) Cl, P, Si
    (B) N, P, As
    (C) Sr, Ca, Mg
    (D) Cl, Br, I
    (E) F, Ne, Na

7. The overall electron configuration of the sulfide ion is most similar to (isoelectronic with) the electron configuration of the
    (A) oxide ion
    (B) chlorine atom
    (C) oxygen atom
    (D) sodium ion
    (E) potassium ion

8. Which describes the behavior of potassium metal during a chemical reaction?

       I.   Neutral atoms become ions with a 1+ charge.
      II.   Neutral atoms take on protons.
    III.   Neutral atoms become ions with a corresponding increase in radius.

   (A) I only
   (B) I and II only
   (C) III only
   (D) II and III only
   (E) I, II, and III

9. Of the elements listed, which is the heaviest element whose atoms have more s electrons than p electrons?

   (A) $_5$B
   (B) $_7$N
   (C) $_9$F
   (D) $_{12}$Mg
   (E) $_{13}$Al

10. The value of Planck's constant is 6.63×10^{-34} J sec. The velocity of light is 3.0×10^8 m sec^{-1}. Which value in joules is closest to the energy of a photon with frequency of 8.0×10^{15} sec^{-1}?

    (A) 1×10^{-20}
    (B) 5×10^{-20}
    (C) 5×10^{-19}
    (D) 1×10^{-18}
    (E) 5×10^{-18}

11. The value of Planck's constant is 6.63×10^{-34} J sec. The speed of light is 3.0×10^{17} nm sec^{-1}. Which value is closest to the wave length in nanometers of a quantum of light with frequency of 6×10^{15} sec^{-1}?

    (A) 10
    (B) 25
    (C) 50
    (D) 75
    (E) 100

12. What is the number of half-filled orbitals in an atom of phosphorus?

    (A) none
    (B) one
    (C) three
    (D) five
    (E) seven

13. Which description of electron configuration applies to atoms of potassium, chromium and copper?

   (A) only one half-filled orbital
   (B) only one filled sublevel
   (C) only three half-filled orbitals
   (D) only nine filled orbitals
   (E) only five filled sublevels

14. How does the electron structure of a phosphorus atom differ from that of a phosphide ion?

   I.   The phosphide ion has more electrons.
   II. The phosphorus atom has more unpaired electrons.
   III. The phosphide ion has more kernel (core) electrons.

   (A) I only
   (B) I and II only
   (C) I and III only
   (D) II and III only
   (E) I, II, and III

15. What is the number of electrons in an atom of $_{23}$V that have an ℓ quantum number of 2?

   (A) 2
   (B) 3
   (C) 6
   (D) 10
   (E) 12

16. The electron configuration of atoms of germanium, $_{32}$Ge, is shown below:

$$[\text{Ar core}] \; 3d^{10} \, 4s^2 \, 4p^2$$

   This element is known to form more than one oxide. Which pair of formulas includes the two most likely formulas for these oxides?

   (A) GeO and GeO_2
   (B) GeO_2 and Ge_2O_7
   (C) Ge_2O_3 and GeO_2
   (D) GeO and Ge_2O_3
   (E) Ge_2O_3 and Ge_2O_7

17. The most abundant isotopes of hydrogen and oxygen are $_1^1H$, $_1^2H$, $_8^{16}O$, and $_8^{17}O$, respectively. Using these isotopes only, what is the number of different possible values for the molar mass of water in grams?

    (A) 2
    (B) 3
    (C) 4
    (D) 6
    (E) 8

18. Which set of formulas or symbols best illustrates the Law of Multiple Proportions?

    (A) $_6^{12}C$ and $_6^{14}C$

    (B) $_6^{14}C$ and $_7^{14}N$

    (C) CO and CO_2

    (D) CH_4 and CCl_4

    (E) C_2H_5OH and CH_3OCH_3

19. All of the following can be inferred from the Lewis dot diagram of a neutral atom below EXCEPT

    (A) X belongs to the same family as sulfur.
    (B) X has two half filled p-orbitals.
    (C) X has at least ten kernel (core) electrons.
    (D) X can accept two electrons to become an ion with a charge of 2^-.
    (E) X has at least four electrons with ℓ quantum number of 1.

20. Consider the set of quantum numbers: 3, 2, -1, $-\frac{1}{2}$

    Which set of quantum numbers represents an electron with the same energy but different orientation in space as the electron represented above?

    (A) 3, 2, -1, $+\frac{1}{2}$

    (B) 3, 1, -1, $-\frac{1}{2}$

    (C) 3, 2, 0, $+\frac{1}{2}$

    (D) 2, 1, 0, $+\frac{1}{2}$

    (E) 2, 2, -1, $-\frac{1}{2}$

21. What is the number of filled orbitals in a ground state atom of manganese, $_{25}$Mn?

    (A) 7
    (B) 10
    (C) 12
    (D) 13
    (E) 15

22. Which gives a list of species with the same electron configuration; that is, species that are isoelectronic?

    (A) Mg, Ca, Sr

    (B) Mg^{2+}, Ca^{2+}, Sr^{2+}

    (C) F^-, S^{2-}, As^{3-}

    (D) Fe, Co, Ni

    (E) S^{2-}, Cl^-, K^+

23. Which color in the visible spectrum is associated with the lowest frequency?

    (A) blue
    (B) green
    (C) red
    (D) violet
    (E) yellow

24. According to quantum mechanics, what is the maximum number of electrons that can occupy the third energy level in a ground state atom?

    (A) 2
    (B) 4
    (C) 8
    (D) 18
    (E) 32

25. Which pair of atoms represents nuclei that have the same number of neutrons?

    (A) ^{56}Co and ^{58}Co

    (B) ^{57}Mn and ^{57}Fe

    (C) ^{58}Ni and ^{57}Fe

    (D) ^{58}Ni and ^{57}Co

    (E) ^{59}Ni and ^{56}Fe

Free-Response Questions

26. Account for each of the following in terms of models of atomic structure.

    (A) In aqueous solution, Ti^{3+} is colored but Sc^{3+} is colorless.
    (B) Chlorine is a better oxidizing agent than iodine.
    (C) Barium is a better reducing agent than magnesium.
    (D) Both the sulfide ion and the chloride ion have 18 electrons. However, the radius of the sulfide ion is greater.

27. Use principles of atomic structure to answer the following questions:

    (A) Write the electron configuration and the Lewis structure for a ground state atom of arsenic.

    (B) The first ionization energy of arsenic is greater than the first ionization energy of germanium. Explain.

    (C) The second ionization energy of arsenic is greater than the first ionization energy of arsenic. Explain.

    (D) When an arsenic atom becomes an arsenide ion as in Na_3As, it gains three electrons. Write the set of four quantum numbers for any two of these three added electrons.

28. Consider the distribution of electrons in the orbitals of a ground-state germanium atom as shown below:

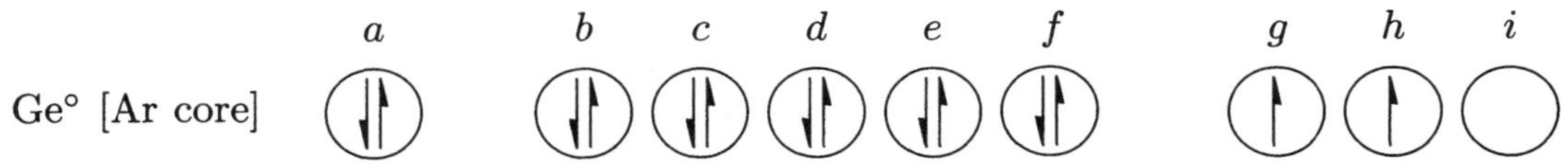

    (A) Refer to orbital a.

        (1) Specify a set of quantum numbers that can be used to represent one of the electrons in orbital a.

        (2) Will the set of quantum numbers that represents the second electron in orbital a be different? Explain.

        (3) What is the maximum number of electrons that can occupy the principal energy level that contains orbital a?

    (B) Refer to orbitals b, c, d, e, and f.

        (1) What is the number of sublevels in the principal energy level that includes orbital d?

        (2) What is the range of signs and values for the m (m_ℓ) quantum numbers for electrons in orbitals b, c, d, e, and f?

    (C) What is the range of signs and values for the ℓ quantum numbers in orbitals g, h, i?

    (D) From which orbitals will electrons be removed when the Ge° atom is converted to the Ge^{4+} ion?

CHAPTER 2
CHEMICAL BONDING

Chemical bonding and atomic structure are major topics in Advanced Placement Chemistry. You should be ready to answer with confidence at least one free response question on this topic on every exam. You will encounter these principles will often in the multiple choice section of the exam.

Chemical bonds are the forces that hold atoms together. Atoms form bonds in order to attain a minimal energy state. Bond formation is an exothermic process (just as bond breaking is an endothermic process). The type and strength of bond that forms between reacting particles dictates the physical and chemical characteristics of the molecule or ion in question.

CHEMICAL BONDS: COVALENT AND IONIC, OTHER

Ionic bonds occur between ions due to electrostatic attraction between positive cations and negative anions. These bonds form between atoms with large differences in electronegativity (> 1.5, using the Pauling scale), usually between a metal and a non-metal. The relative strength of an ionic bond can be inferred using Coulomb's Law. The strength of an ionic bond is directly proportional to the magnitude of the charges involved and inversely proportional to the square of the distance between them. Ionic solids tend to have high melting points and are often soluble in polar solvents, such as water.

Coulomb's Law can be used to predict the relative attraction between ions. This information helps predict melting point of ionic compounds.

$$E_{coulomb} = \frac{k(q_1)(q_2)}{r^2}$$

where q_1 = magnitude of charge on one ion, q_2 = magnitude of charge on the other oppositely charged ion, and r = sum of ionic radii.

Nonpolar covalent bonds form between atoms of nonmetals with nearly identical electronegativities while polar covalent bonds form between nonmetals with dissimilar electronegativity values. While covalent bonds within molecules are strong, the binding forces between molecules are often relatively weak. (See *intermolecular forces* below.) Molecular solids tend to have low melting points and may be soluble in non-polar solvents such as tetrachloromethane (carbon tetrachloride).

Metallic bonding occurs in metallic solids. Metal atoms tend to have a large positive charge on the nucleus with relatively few valence electrons. Those nuclei are positioned in a regular geometric array (lattice) by electrostatic repulsion while their valence electrons as a diffuse cloud of electrons are equally attracted by adjacent nuclei. This leads to the "sea of electrons" model, wherein the nuclei bob like islands in a sea of free-flowing electrons. This model is useful for explaining physical characteristics such as electrical and thermal conductivity. Metals have a wide range of melting points.

Intermolecular forces (IMF) is applied to the group of weaker attractive forces between atoms or molecules that are not classified as bonds. Known generically as "**van der Waals forces**", they include three types. London dispersion forces are common to all atoms and molecules. These are caused by temporary dipoles that occur when electrons shift around the nuclei, thus forming ephemeral attractions and repulsions. Although these are very weak, they exist between all atoms and molecules. The strength of the temporary dipole formed depends on the number of electrons that are moving around the molecule. Molecules with larger molecular mass (and therefore more electrons) have greater London forces. Another general sort of IMF is the dipole-dipole force, an attraction between the opposite polar ends of adjacent molecules. The third category of van der Waals forces is an exaggerated form of the dipole force that occurs when the electropositive hydrogen atom bonds with a very electronegative partner, such as fluorine, oxygen, or nitrogen. The "H-FON" bond formed is thus very polar; the strongly positive and negative ends of adjacent molecules have noticeably stronger interaction than a simple London force. This is called **hydrogen bonding**.

LEWIS STRUCTURES

Lewis structures are drawn with a technique that is based on grouping valence electrons in octets around each nucleus. Chemical bonds are often represented as dashes – one dash represents one shared electron pair. A pair of dots (or other symbol) can represent a shared or unshared electron pair. Square brackets enclose an ion with its charge placed at the upper right. Drawing and interpreting Lewis structures occurs on every exam. Some examples of Lewis structures that illustrate the octet rule given in Figure 2.1. Some exceptions are shown in Figure 2.2. The technique can be extended to groupings of atoms to form molecules or polyatomic ions. The positioning of electron pairs around a central atom can be predicted using a technique known as Valence Shell Electron Pair Repulsion Theory (VSPERT). Based on location of electron pair, you can then predict the positioning of atoms within a molecule or polyatomic ion. The polarity of the predicted molecule can be used to predict the strength of the intermolecular forces as well as resulting physical characteristics such as vapor pressures and melting and boiling points.

Figure 2.1 Selected Lewis Structures illustrating the octet rule

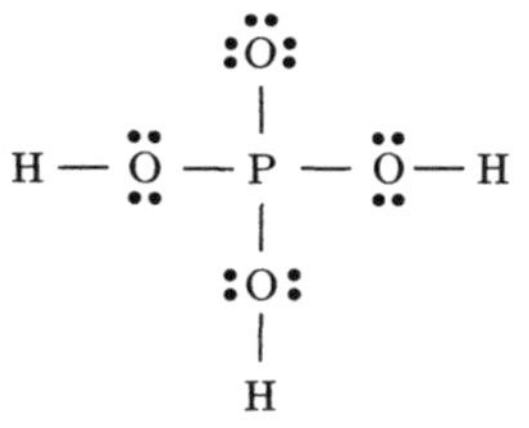

H_3PO_4 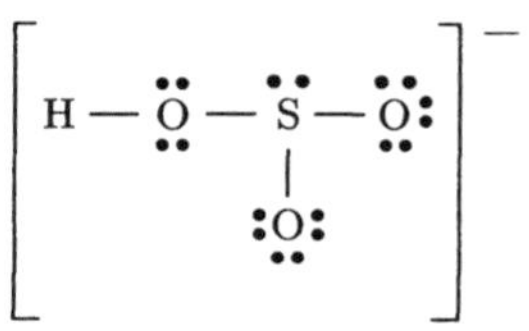 HSO_3^- 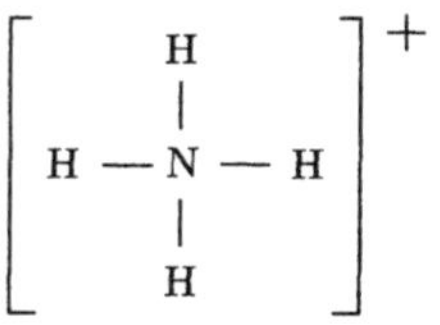 NH_4^+

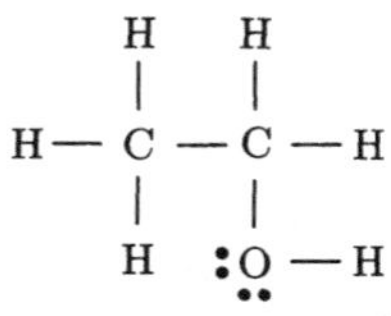

C_2H_5OH 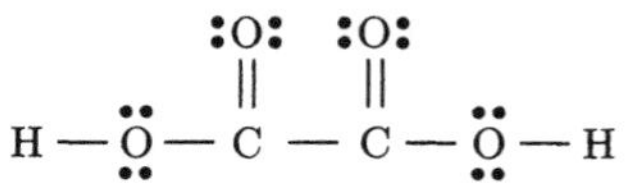 $H_2C_2O_4$ 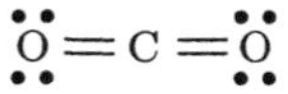 CO_2

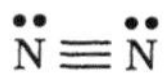

N_2 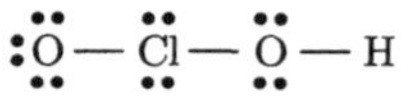 $HClO_2$

Figure 2.2 Selected Lewis Structures: exceptions to the customary format

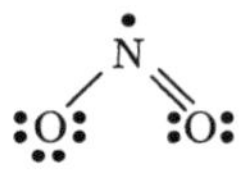

NO_2, an odd-electron molecule

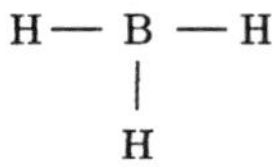

BH_3, three (rather than four) shared electron pairs

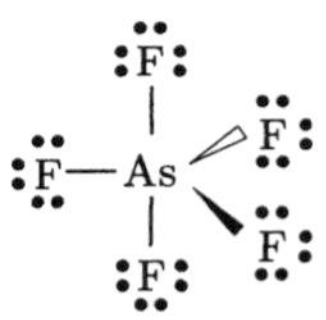

AsF_5, expanded octet, five shared electron pairs

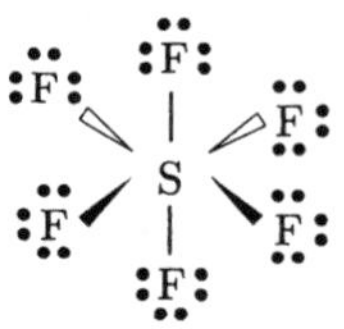

SF_6, expanded octet, six shared electron pairs

VOCABULARY: BONDS AND BONDING

Covalent bond	bond formed from shared electrons.
Bond length	average distance between nuclei of two bonded atoms in a molecule.
Bond energy	energy needed to break one particular chemical bond in a gaseous substance.
Lewis Structures	representation of molecules and atoms using symbols for atoms with dots and lines for electrons.
Bond energy	a measure of the stability of a chemical bond; the amount of energy needed to break the bond. This depends on the atoms that form the bond specified and on the effects of all the other atoms in the molecular structure. Bond energy is also determined by the number of electrons shared; bond strength increases with bond multiplicity. It also increases with increase in electronegativity difference, due to increased stability from coulombic attractions. Bond strength decreases with decreased orbital overlap between atoms. Because those valence electrons are more diffuse and electron density is more spread out, overlap on the axis and resulting attraction decreases.
Bond length	characteristic distance of separation between two atoms, influenced by principal quantum number, because a greater number of core electrons decreases the ability of two atoms to get very close to each other due to electron repulsion. Increased multiplicity of a bond decreases bond length because the placement of additional electrons between atoms increases coulombic attraction. Higher effective nuclear charge (Z_{eff}) decreases bond length because higher results in smaller atomic size, allowing the atoms to get closer to each other. Increased electronegativity decreases bond length because the partial charges increase the coulombic attraction, allowing the atoms to get closer. The ability to form a *pi* bond is dependent on atomic size. For example C=C is possible due to the relatively small atomic size, but Si=Si is not because the Si atoms are much larger. Breaking a bond *always* requires energy; forming a bond always releases energy.
Molecular orbitals	When two atomic orbitals from different atoms interact, two new molecular orbitals are generated, one additive and one subtractive. The additive orbital (**bonding molecular orbital**) has high electron density between the nuclei. The subtractive orbital (**antibonding molecular orbital**) has low electron density between the nuclei. The bonding and antibonding σ_s orbitals are more stable than any molecular orbital from $2p$, because $2s$, which forms σ_s and $\sigma_s{}^*$, is more stable that $2p$. The two π bonding orbitals are of equal energies, as are the π^* orbitals. The $2p$ antibonding orbitals are the least stable molecular orbitals, with $\sigma_p{}^*$ less stable than π^*.
Bond order:	net amount of bonding between two atoms, $$BO = \tfrac{1}{2} \text{ the number of } (e^-{}_{bonding} - e^-{}_{antibonding})$$

Bonding orbitals	orbital with high electron density between the atoms
Sigma (σ) **bond**	bond formed by the end-on overlap of atomic orbitals giving high electron density along the axis, with one pair of electrons shared between two atoms
Pi (π) **bond**	bond formed by side-by-side overlap of atomic orbitals accounting for electron density above and below the axis
Double ($\sigma + \pi$) **bond**	chemical bond containing two pairs of electrons, involving two orbital orientations, with one pair end-on-end and a second pair side-by-side
Triple ($\sigma +$ two π) **bond**	chemical bond with three shared pairs of electrons, involving three orientations, one end-on and two side-by-side
Nonbonding electrons	valence electrons that do not participate in bonding
Lone pairs	a pair of valence electrons that is localized on one atom instead of being involved in a chemical bond
Formal charge (FC)	"bookkeeping" of valence electrons in Lewis structures; calculate the apparent charge on any atom in a Lewis structure as follows: • all unshared electrons assigned to atom where located • half the bonding electrons in a shared pair assigned to each of the bonded atoms The formal charge for each atom in the Lewis structure is equal to the number of valence electrons in that isolated atom minus the number of electrons assigned as above to that atom in the Lewis structure.
Resonance structures	two or more equivalent Lewis structures representing a molecule with delocalized electrons
Delocalized π orbitals	formed when more than two p orbitals from three or more atoms overlap in the appropriate geometry. Evidence for the orbitals is provided by bond lengths: in ozone, O_3, all bond lengths are equal and intermediate between that of $O-O$ and $O=O$. Delocalized π orbitals affect absorption spectra/color, bond stability, and redox. For example, many organic molecules with these orbitals are colored (eg. chlorophyll) .
Polar covalent bond	asymmetric electron distribution between two bonded nuclei
Electronegativity	measure of an atom's ability to attract shared electrons in a bond
Dipole moment	the net electrical character arising from asymmetrical charge distribution in a molecule or polyatomic ion

VOCABULARY: MOLECULAR AND ELECTRON PAIR GEOMETRY

VSEPR	principle of minimizing electron-electron repulsion by placing electron pairs as far apart as possible.
Hybridization	formation of a set of hybrid orbitals with favorable directional characteristics by blending two or more valence orbitals of the same atom.
Coordination structure	the number of atoms to which an atom is bonded.
Steric number	the sum of the coordination number of an atom and its number of lone pairs of electrons.

Molecular shapes

(a) **tetrahedron**	electron or molecular geometry that features four identical faces that are equilateral triangles.
trigonal pyramid	pyramid shape with an equilateral triangle base and isosceles triangle sides due to a lone pair distorting three shared pairs of electrons and the corresponding bond angles.
bent shape	shape of a triatomic molecule with a bond angle $< 180°$, due to two lone pairs.
(b) **trigonal bipyramidal**	double pyramid with triangular base and two apices along a linear axis perpendicular to the plane of the base, dsp^3 hybrid.
trigonal bipyramid	five atoms arranged around central atom, with bond angles of $90°$ and $120°$.
seesaw shape	four outer atoms and one central atom; one equatorial lone pair; bond angles of $< 120°$ and $< 90°$ due to lone pair.
(c) **octahedral shape**	double pyramid with square base; six vertices for atoms and/or lone pairs, d^2sp^3 hybridization.
octahedron	six outer atoms, with bond angles $= 90°$.
square planar	four outer atoms and two axial lone pairs, bond angles $= 90°$.
square pyramid	five outer atoms and one axial lone pair, bond angles $< 90°$.
(d) **trigonal planar**	sp^2 hybrid with three atoms arranged to form a triangle bond angle $= 120°$.
linear geometry	three atoms arranged in a line.

MOLECULAR GEOMETRY

The best way to address the concept of molecular geometry is to consider first the distribution of electron pairs. Then consider the location of shared and unshared (lone pairs) of electrons. In general, pairs of electrons are distributed to minimize repulsion. Unshared pairs are located so as to be as from far from each other as possible. A systematic presentation of electron pair distribution is given below. Consult your textbook for detailed diagrams that include representation of three dimensions.

Covalent bonds; electron pair distribution and hybridization

- Two pairs: linear distribution; two sp hybrid orbitals

shared pairs	$-A-$
two	linear – three atoms arranged in a straight line

- Three pairs: trigonal planar distribution; three sp^2 hybrid orbitals

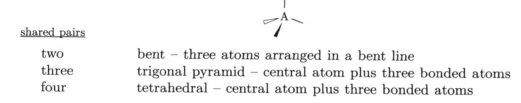

shared pairs	
two	bent – three atoms arranged in a bent line
three	trigonal planar – central atom plus three bonded atoms

- Four pairs: tetrahedral distribution; four sp^3 hybrid orbitals

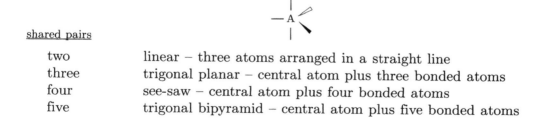

shared pairs	
two	bent – three atoms arranged in a bent line
three	trigonal pyramid – central atom plus three bonded atoms
four	tetrahedral – central atom plus three bonded atoms

- Five pairs: trigonal bipyramidal distribution; five dsp^3 hybrid orbitals

shared pairs	
two	linear – three atoms arranged in a straight line
three	trigonal planar – central atom plus three bonded atoms
four	see-saw – central atom plus four bonded atoms
five	trigonal bipyramid – central atom plus five bonded atoms

- Six pairs: octahedral distribution; six d^2sp^3 hybrid orbitals

shared pairs	
two	linear – three atoms arranged in a straight line
three	T-shaped – central atom plus three bonded atoms
four	square planar – central atom plus four bonded atoms
five	square pyramid – central atom plus five bonded atoms
six	octahedron – central atom plus six bonded atoms

A summary of distribution of electron pairs and resulting molecular geometry is found in Figure 2.3.

Figure 2.3 Distribution of electron pairs

Electron pairs	Geometrical arrangement	Electron-pairs (*sigma* bonds)		Set of hybrid orbitals	Molecular shape	Examples
		shared	unshared			
2	linear	2	0	sp	linear	$BeCl_2$; CO_2
3	trigonal planar	3	0	sp^2	trigonal planar	BF_3; SO_3
		2	1		bent	SO_2
4	tetrahedral	4	0	sp^3	tetrahedral	CH_4
		3	1		trigonal pyramidal	NH_3; SO_3^{2-}
		2	2		bent	H_2O
		1	3		linear	HF
5	trigonal bipyramidal	5	0	dsp^3	trigonal bipyramidal	PCl_5
		4	1		see-saw	SF_4
		3	2		T-shaped	ClF_3
		2	3		linear	XeF_2
6	octahedral	6	0	d^2sp^3	octahedral	SF_6; $Fe(CN)_6^{3-}$
		5	1		square pyramidal	BrF_5
		4	2		square planar	XeF_4; $Cu(H_2O)_4^{2+}$

from the TOPIC OUTLINE (website: apcentral.collegeboard.com)

I. Structure of Matter

B. Chemical Bonding

   1. Binding forces

      a. Types: ionic, covalent, metallic, hydrogen bonding, van der Waals (including London dispersion forces)

      b. Relationships to states, structure, and properties of matter

      c. Polarity of bonds, electronegativities

   2. Molecular models

      a. Lewis structures

      b. Valence bond: hybridization of orbitals, resonance, sigma and pi bonds

      c. VSEPR

   3. Geometry of molecules and ions, structural isomerism of simple organic molecules and coordination complexes; dipole moments of molecules; relation of properties to structure

from the list of CHEMICAL CALCULATIONS

No specified chemical calculations are closely associated with chemical bonding. However, use of bond energies does occur.

from the list of EQUATIONS & CONSTANTS

None of the entries in this list is closely associated with Chemical Bonding

from the list of RECOMMENDED EXPERIMENTS

From the laboratory list:

  1. Determination of the formula of a compound

  2. Determination of the percentage of water in a hydrate

 15. Synthesis of a coordination compound and its chemical analysis

 22. Synthesis, purification, and analysis of an organic compound

Multiple Choice Questions

Questions 1-5: The set of lettered choices below is a list of classes of solids and refers to the numbered phrases immediately following it. Select the one lettered choice that best fits each phrase. A choice may be used once, more than once or not at all.

(A) an ionic solid
(B) a metallic solid
(C) a network solid with covalent bonds
(D) a molecular solid with hydrogen bonding
(E) a molecular solid with nonpolar molecules

1.   Cu, copper wire

2. I_2, iodine crystals

3. $C_{12}H_{22}O_{11}$, granular sugar

4. $MgSO_4$, magnesium sulfate crystals

5.   SiC, powdered silicon carbide

------

6.   All of these molecular shapes can be explained by dsp^3 hybridization of electrons on the central atom EXCEPT

(A) linear
(B) T-shape
(C) see-saw
(D) octahedral
(E) trigonal bipyramid

7.   Which of the following molecules is predicted to have the greatest molecular dipole moment?

(A) CO_2
(B) HBr
(C) HCl
(D) HI
(E) O_2

8.   All species below have Lewis dot diagrams that illustrate the octet rule EXCEPT

(A) NO_3^-
(B) NH_3
(C) NH_4^+
(D) N_2
(E) NO_2

9. Which species exhibits molecular geometry usually described as bent?

    (A) HF
    (B) NH_3
    (C) CH_4
    (D) NO_2^-
    (E) NO_3^-

Questions 10-12: Consider the chemical bonds found in solid sodium hydrogen carbonate. For each bond specified, choose the best description from the list of bond types below.

    (A) ionic bond
    (B) hydrogen bond
    (C) single covalent bond
    (D) double covalent bond
    (E) resonance covalent bond with bond order between 1 and 2

10. carbon/oxygen bond

11. sodium/hydrogen carbonate bond

12. oxygen/hydrogen bond

---

13. In the sulfur dioxide molecule, the O-S-O bond angle is slightly less than 120°. Which distribution of electrons around the central atom, sulfur, provides the best explanation for this bond angle?

	unshared electron pairs	shared electron pairs	shared resonance electron pairs
(A)	0	2	1
(B)	0	2	2
(C)	1	2	0
(D)	1	0	2
(E)	1	2	1

14. The H-C-OH bond angle in methanoic acid (formic acid), HCOOH, is slightly less than 116°. Which distribution of electron pairs **around the carbon atom** provides the best explanation for this bond angle?

	unshared electron pairs	shared electron pairs in single bonds	shared electron pairs in double bonds
(A)	0	2	2
(B)	0	1	2
(C)	1	2	1
(D)	2	2	0
(E)	3	2	2

15. Which substance has the greatest ionic character?

   (A) Cl_2O
   (B) NCl_3
   (C) $PbCl_2$
   (D) $MgCl_2$
   (E) CH_2Cl_2

16. The shape of the IO_3^- ion is best described as

   (A) see-saw
   (B) T-shaped
   (C) tetrahedral
   (D) trigonal planar
   (E) trigonal pyramidal

17. Which gives the correct comparison of the carbon-carbon bond characteristics in C_2H_2 and C_2H_4?

   Compared to C_2H_4, the property listed for C_2H_2 is

	bond length	bond strength	number of shared electron pairs
(A)	greater	smaller	greater
(B)	greater	smaller	smaller
(C)	smaller	smaller	smaller
(D)	smaller	greater	greater
(E)	greater	greater	greater

18. Which pair of characteristics is most closely associated with metallic solids?

      I.   low melting point

     II.   high malleability

   III.   low thermal conductivity

   IV.   high electrical conductivity

(A) I and II
(B) I and III
(C) II and III
(D) II and IV
(E) III and IV

19. Which of the following identifies the species that occupies the lattice points in crystals of xenon and xenon tetrafluoride?

(A) Some lattice points in each substance contain anions.

(B) Some lattice points in each substance contain cations.

(C) The lattice points in xenon are occupied by atoms while the lattice points in xenon tetrafluoride are occupied by molecules.

(D) The lattice points in xenon are occupied by ions while the lattice points in xenon tetrafluoride are occupied by atoms.

(E) The lattice points in xenon are occupied by ions while the lattice points in xenon tetrafluoride are occupied by molecules.

20. Consider the molecular geometries:
    linear, bent, T-shaped, trigonal planar, trigonal pyramidal
Each of the following species is described by one of the geometries above EXCEPT

(A) BeF_2
(B) BF_3
(C) CCl_4
(D) BrI_3
(E) H_2O

21. The PF_6^- ion is known to have octahedral geometry. Explanation of its bonding includes all of the following EXCEPT

(A) bond angles of 90°
(B) resonance structures
(C) expanded octet
(D) d^2sp^3 hybridization
(E) electrons shared as single bonds

22. Which identifies two ions, from the list below, that have geometry that is explained by the same hybridization of orbitals?

$$NO_2^-, NO_3^-, NH_4^+, SCN^-, NH_2^-$$

   (A) NO_2^- and NO_3^-

   (B) NO_2^- and SCN^-

   (C) NH_4^+ and NO_3^-

   (D) SCN^- and NH_2^-

   (E) NO_2^- and NH_2^-

23. Which correctly compares single bonds with equal sharing of electrons to single bonds with unequal sharing of electrons?

   I.   Bonds with equal sharing are weaker.

   II. Bonds with equal sharing have smaller bond energy.

   III. Bonds with equal sharing are associated with smaller electronegativity difference between atoms.

   (A) I only
   (B) III only
   (C) I and II only
   (D) I and III only
   (E) I, II, and III

24. What is the hybridization of orbitals in the ammonia molecule, NH_3?

   (A) sp
   (B) sp^2
   (C) sp^3
   (D) dsp^3
   (E) d^2sp^3

25. Consider a molecule of 1,3-butadiene (C_4H_6). Which identifies the correct numbers of σ and π bonds in the molecule?

   (A) $1\sigma, 2\pi$
   (B) $6\sigma, 3\pi$
   (C) $7\sigma, 2\pi$
   (D) $7\sigma, 4\pi$
   (E) $9\sigma, 2\pi$

Free-Response Questions

26. Using principles of chemical bonding and/or intermolecular forces, explain each of the following.

    (A) The normal boiling point of iodine, I_2, is greater than the normal boiling point of chlorine, Cl_2.

    (B) Both $Ag_{(s)}$ and molten Ag are excellent conductors of electricity. However, silver nitrate, $AgNO_3$, is a good conductor only when melted or dissolved in water. As a solid, it is a poor conductor of electricity.

    (C) The normal boiling point of H_2O is higher than the normal boiling point of H_2S even though the molar mass of H_2O is less than the molar mass of H_2S.

    (D) Arsenic, As, reacts with the metal sodium, Na, to form Na_3As. Arsenic reacts with the nonmetal chlorine, Cl_2, to form $AsCl_3$.

27. The shape of the carbonate ion, $CO_3{}^{2-}$, is known to be trigonal planar (triangular coplanar). Each of the C-O bonds is known to be equivalent.

    (A) Draw a Lewis electron dot diagram for any one of the three equivalent contributing resonance structures of the carbonate ion, $CO_3{}^{2-}$.

    (B) Discuss the hybridization of the atomic orbitals on the central carbon atom that is most likely to account for the bonding, geometry and distribution of electrons within this polyatomic ion. Include the role of *sigma* (σ) and *pi* (π) bonds in determining geometry and electron distribution.

    (C) Draw a reasonable 3-dimensional representation of this ion. Clearly label *sigma* and *pi* bonds.

    (D) Using a chemical symbol or formula, identify the atom, ion or molecule found at each lattice point in the solid crystalline form of calcium carbonate.

28. Although the number of atoms and overall charge of the azide ($N_3{}^-$) and triiodide ($I_3{}^-$) ions are similar, the bonding in the two polyatomic ions is quite different.

    (A) Draw a Lewis structure for each of the ions. Indicate the electron pair geometry, molecular geometry, and hybridization that applies to the central atom for each ion.

    (B) Which, if any, of the ions is polar? Explain.

    (C) Which, if any, of the ions has delocalized π bonds? Explain.

CHAPTER 3
THE PHASES OF MATTER – SOLID, LIQUID, AND GAS

THE KINETIC MOLECULAR THEORY

Which came first? Beginning in the late 16th century, systematic controlled investigation of the natural world established the Gas Laws. By the early 19th century, enough was known about how gases behave, such that scientists could propose a theory to account for this behavior. The 21st century version of this theory is called the Kinetic Molecular Theory (KMT).

That phrase should trigger a mental picture of particles in rapid random motion, colliding with each other and the walls of the container. These particles never stick to each other or to the walls of the container. They never lose any energy; they just keep on moving.

You should be able to cite and use appropriately some ideas which many authorities call the postulates of the Kinetic Molecular Theory.

Gases are composed of particles (atoms or molecules) that

- are in constant random, rapid motion

- are nearly an infinite distance apart, compared to the size of the particle

- have no dimensions and take up no space

- experience no forces of attraction or repulsion for each other

- collide with each other and the wall of the container with no loss of energy

- have average kinetic energy that is proportional to the absolute temperature of the system

The gases that "obey these rules" are called ideal gases.

What's energy, and especially average kinetic energy? Principles of physics give the definition of the kinetic energy of a body in motion (kinetic means related to motion) as equal to $\frac{1}{2}mv^2$. In words, this means that a particle with mass, m in kg, traveling at a velocity, v in m/sec, exhibits a phenomenon we call energy, and measure with a thermometer as the absolute (Kelvin) temperature that is proportional to $\frac{1}{2}mv^2$.

Thus, the average kinetic energy of a very large number of particles (molecules) is what we measure with a thermometer.

What's an ideal gas? ... or a non-ideal gas?

An ideal gas is a gas that illustrates the postulates of the Kinetic Molecular Theory over a wide range of temperatures or at least over the temperature range in question. In most gas law problems, it is assumed that the gas behaves "ideally". When gases do not behave ideally, they are often said to be "real gases". In general, most gases become nonideal when they reach conditions approaching the liquid phase.

This theory isn't "true". Everyone knows that if enough pressure is applied to a system maintained at a low enough temperature, almost any gas will change to a liquid and that the particles no longer move from place to place. When these particles move slow enough and get close enough to each other, their forces of attraction take over. They stick together. Furthermore, like all atoms and molecules, these particles do have dimension and do take up space.

Once a gas changes to a liquid, the gas laws no longer apply and the kinetic molecular theory for gases needs to be revised. But for nearly any system that is a gas, the KMT does a very good job of predicting behavior for most gases.

THE GAS LAWS (ACTUALLY THE GAS OBSERVATIONS)

You probably will never earn credit for simply naming a gas law. But knowing the name may help you organize a few ideas and a few corresponding experiments.

NAMING THE GAS LAWS

Boyle's Law	pressure/volume relationship for a sample (constant number of molecules) held at constant temperature; push down on a piston, get a smaller volume
Charles's Law	temperature/volume relationship for a sample (constant number of molecules) held at constant pressure; heat a gas-filled piston, get a larger volume
"Balloon's" Law	(see Avogadro's Hypothesis below) number of particles/volume relationship; blow into a balloon, get a larger balloon.
Dalton's Law	for a mixture of gases, add their partial pressures to get total pressure
Gay-Lussac's Law	when gases react chemically, the ratios of their reacting volumes are small integers
Graham's Law	at the same temperature, heavy molecules move slower than light molecules (find the equation on the *List....*)
Avogadro's hypothesis	offers the explanation that equal volumes of gases at the same temperature and pressure have equal numbers of molecules. That accounts for Gay-Lussac's Law and helps account for Dalton's Law.

Figure 3.1 Boyle's Law

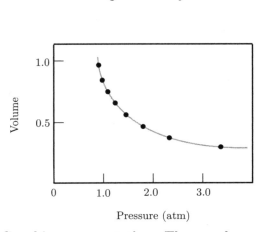

Pressure (atm)

Figure 3.2 Charles's Law

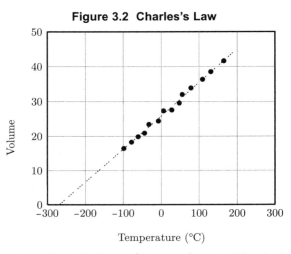

Temperature (°C)

Graphic representations The graphs representing Boyle's Law (Fig. 3.1) and Charles's Law (Fig. 3.2) are often used in examinations.

MATHEMATICS OF THE GENERAL GAS LAW

The Kinetic Molecular Theory accounts for all these observations. Its mathematical expression is called the general gas law (or ideal gas law),

$$PV = nRT$$

where P is pressure in atm, V is volume in L, n is number of particles in moles, and T is temperature in kelvins, the universal gas constant R is 0.0821 L atm mol^{-1} K^{-1}.

This equation is found in the *List...* with other combinations of appropriate labels. This equation must become part of your storehouse of personal knowledge.

CALCULATIONS USING THE GAS LAWS

Density Density is always expressed in units of mass per unit volume. The formula given in the *List...* is best described as a definition. The values for density of common gases at ordinary conditions vary from about 0.10 g L^{-1} for hydrogen to over 3.0 g L^{-1} for chlorine.

The Molar Volume If equal volumes of gases contain equal numbers of molecules, then one mole of any ideal gas occupies the same volume as any other ideal gas. Experiments show that one mole of any gas occupies 22.4 L at STP– one atmosphere of pressure and 273 kelvins of temperature (that is, of average kinetic energy).

The four variables system:

Pressure, volume, temperature, number of moles of molecules

1. Given three of the four variables directly, find the fourth by using the general gas law.
    $$PV = nRT$$
    Be sure to use consistent units for all four variables and the gas constant, R.

2. Given two of the four properties held constant, find the value of the fourth variable when the third variable changes.

3. Given three of the four variables, perhaps somewhat indirectly or not-so-directly, find the fourth by using the general gas law. When the mass of the sample is given and the formula mass is known, the number of moles of molecules can be determined, then used as one of the three variables necessary to determine the fourth.

Using density

4. When density is known in grams per liter, note that two facts are known about the sample. A one-liter sample has mass that has been specified. The molar mass or number of moles of molecules can be immediately determined using the principle that one mole of molecules of an ideal gas occupies 22.4 liters at STP. If the density is reported at some conditions that are not STP, use the principles of Boyle's and Charles's Laws to change that one-liter volume to a new volume, hence a new density, at STP.

Mixtures of gases

5. In some systems, two or more gases are present in the same sample. All the components of the mixture have the same volume and the same temperature. However, each does not necessarily have the same number of particles or the same (partial) pressure. The effects of the collisions of all the molecules add together to give the total pressure. Thus, Dalton's Law

$$P_{total} = P_A + P_B + P_C + \dots$$

where the total pressure is equal to the sum of the (partial) pressures of each of the component gases, A, B, C, ... is seen to be a very logical illustration of the Kinetic Molecular Theory in action. A common mixture of gases is the gas produced when an alkaline earth metal is added to dilute acid with the resulting production of hydrogen. If the hydrogen is collected by the displacement of water as shown in Figure 3.3, the hydrogen is said to be "wet", that is, saturated with water vapor. The vapor pressure for water in such a sample can be determined from a reference table. Sometimes vapor pressure information is expressed in an equilibrium context such as

$$H_2O_{(\ell)} \rightleftharpoons H_2O_{(g)} \qquad K_p = 0.0313 \text{ atm at } 298 \text{ K}$$

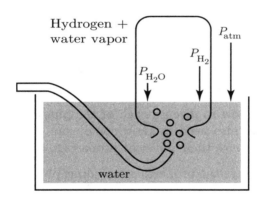

Figure 3.3 Hydrogen collected over water

Molar mass and molecular velocity

6. The most reliable way to deal with questions about molar mass (molecular weight) and molecular velocity is to recognize that these relationships are nearly always expressed with the gases at the same temperature. Thus, these gases have the same average kinetic energy such that

$$\tfrac{1}{2}\, m_A v_A{}^2 = \tfrac{1}{2}\, m_B v_B{}^2$$

where A and B represent two different gases. This relationship can then be rearranged algebraically as needed. When three of the four variables are known, the fourth can be determined algebraically. Sometimes, molecular velocity is described as rate of diffusion with units of distance per unit time. Occasionally, diffusion is presented as time per unit of distance, in which case a greater time indicates a slower rate of diffusion. It is best to avoid trying to memorize all the possible versions of Graham's Law. Use the principle above for every circumstance.

SOLIDS, LIQUIDS AND CHANGES OF PHASE

From the molecular viewpoint

One good way to think about the condensed phases of matter – solids and liquids -- is to simply regard them as gas molecules that have become stuck together. In the gas phase, when molecules collide, they bounce off. If the molecules stick together, they become a liquid. When temperature decreases, molecular velocity decreases. When molecular velocities decrease, forces of attraction become effective and molecules stick together. If they stick together in a fixed physical location, a solid has been formed.

When phase changes occur, both enthalpy and entropy change. Values for changes in enthalpy and entropy can be determined experimentally by using the liquid phase as the starting point. These values are often expressed as enthalpy or entropy of vaporization or solidification. Vaporization means conversion from the liquid to the gas phase, usually at the normal boiling point. Solidification refers to conversion from the liquid to the solid phase, usually at the normal melting point. Lattice energy is one component of enthalpy change in the solidification process. It is equivalent to the energy released when positive and negative ions in the gas phase form an ionic bond in the solid phase as shown in the equation below:

$$Na^+_{(g)} + Cl^-_{(g)} \rightarrow NaCl_{(s)} + energy \text{ (lattice energy)}$$

THE CONDENSED PHASES - SOLIDS AND LIQUIDS

Nearly any kind of atoms, molecules and ions can form a condensed phase - solid or liquid. You need to know how to discuss each of the phases from the Kinetic Molecular perspective, as above. You also need to know some examples of each class of solids and be able to predict and compare their physical properties.

In general, differences in physical properties including phase can be explained in terms of differences in strength and effectiveness of forces of attraction. Compared to substances in the gas phase, liquids and solids have lower vapor pressures, higher melting and boiling points and higher enthalpies of phase change (vaporization/condensation and melting/fusion/solidification). Figure 3.4 provides a useful summary of the thermodynamic characteristics for the phase changes of liquids.

Figure 3.4 Changing Phase of a Liquid

Phase change	ΔH Enthalpy change	ΔS Entropy change	ΔG Free energy change, (starting in the liquid phase)
Vaporization	positive	positive	becomes negative when T is high enough (identifies normal boiling point)
Solidification	negative	negative	becomes negative when T is low enough (identifies normal freezing point)

With the exception of the metals, most other solids and liquids are poor thermal and electrical conductors. Metals are good conductors due to the presence of the diffuse electron cloud associated with metallic bonding. Ionic compounds such as sodium chloride are also good conductors in the liquid phase. However, this is of limited practical application due to the very high melting points of ionic solids.

Figure 3.5 offers a concise summary of information about solids. You should be able to prepare an informed discussion using the facts and principles summarized here.

Figure 3.5 Classification of solids

Class of solid	Particle at the lattice points	Common examples	Forces between particles	Relative strength of forces
Atomic	atoms	argon, krypton	van der Waals	very weak
Molecular	molecules	ice, dry ice, sucrose	van der Waals, dipole/dipole, hydrogen bonding	moderate to strong
Network	atoms	diamond, quartz, graphite	covalent bonds	very strong
Metallic	cations	iron, copper, zinc	metallic bonding	strong to very strong
Ionic	cations and anions	NaCl, $CaCO_3$	ionic bonds	strong

Graphic Representations

Three representations are commonly used to illustrate principles of solids, liquids and phase changes.

Cooling/warming curves are used to track temperature and phase change. Often experiments start in the solid phase with heat added at a constant rate. By observing temperature every minute or similar time period, the observer is able to measure the effect of added heat (at a constant rate) on the temperature of the system. Careful experiments provide data that show two periods of constant temperature, even though heat is being added continuously at constant rate. According to the Kinetic Molecular Theory, when heat is added but no change in temperature occurs, the added heat is used to increase potential energy (change phase) rather than increase kinetic energy (raise temperature). The warming curve for water is shown in Figure 3.6. Note that the time interval for vaporization is larger than the time interval for melting. This means that the heat of vaporization is much greater than the heat of fusion.

Figure 3.6 The warming of water at 1.0 atmosphere

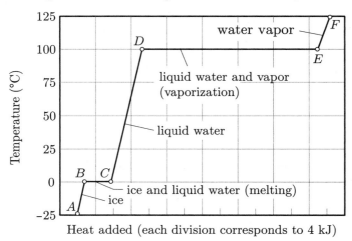

Heat added (each division corresponds to 4 kJ)

<u>Vapor pressure</u> is often presented in a graph format. Figure 3.7 shows the vapor pressure of four substances. Temperature is plotted as the independent variable, with vapor pressure as the dependent variable. The normal boiling point can be determined by reading across the y-value corresponding to 1.0 atmosphere. The temperature at any intersecting line is the normal boiling point for that substance.

Figure 3.7

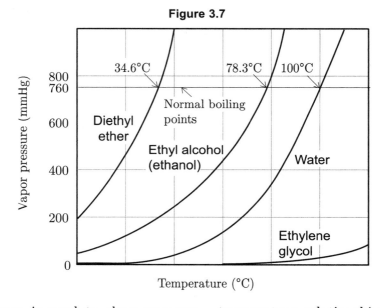

A <u>phase diagram</u> is used to show pressure - temperature relationships. Two simple versions for closed systems of each of two pure substances are found in most introductory chemistry textbooks: one for water (and a few other substances) and another for almost any other substance; carbon dioxide is often used. Figure 3.8(a) shows the phase diagram for water; Figure 3.8(b) shows the phase diagram for carbon dioxide. The solid/gas boundary line indicates the vapor pressure of the solid. The liquid/gas boundary line indicates the vapor pressure of the liquid. The solid/gas boundary line indicates the vapor pressure of the solid. Changes in conditions that occur directly across the liquid/solid boundary line are called melting $((s) \rightarrow (\ell))$ or solidification $((\ell) \rightarrow (s))$. Changes in conditions that occur directly across the liquid/gas boundary line are called vaporization $((\ell) \rightarrow (g))$ or condensation $((g) \rightarrow (\ell))$. Changes in conditions that occur directly across the solid/gas boundary line are called sublimation $((s) \rightarrow (g))$ or deposition $((g) \rightarrow (s))$.

These three lines intersect at the triple point – the temperature at which all three phases are in dynamic equilibrium. For an ideal substance, the solid/liquid line is directly vertical from the triple point. (From a purist point of view, it could be argued that a perfect substance has no triple point because it can exist only as a gas.) In actual practice, the s/ℓ line for water has negative slope while the s/ℓ line for nearly all other substances has positive slope. The negative slope is associated with such anomalous properties as greater density for the liquid phase than the solid phase as in water (ice floats).

Figure 3.8 Phase diagrams

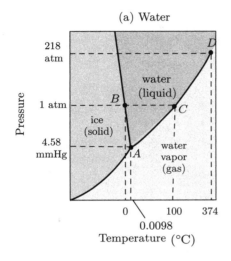

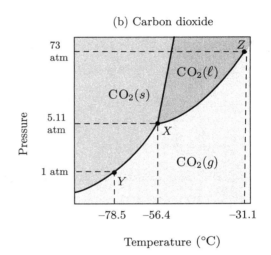

from the TOPIC OUTLINE (website: apcentral.collegeboard.com)

II. States of Matter

A. Gases

    1. Laws of ideal gases

        a. Equation of state for ideal gas

        b. Partial pressures

    2. Kinetic-molecular theory

        a. Interpretation of ideal gas laws on the basis of this theory

        b. Avogadro's hypothesis and the mole concept

        c. Dependence of kinetic energy of molecules on temperature

        d. Deviations from ideal gas laws

B. Liquids and solids

    1. Liquids and solids from the kinetic-molecular viewpoint

    2. Phase diagrams of one-component systems

    3. Changes of state, including critical points and triple points

    4. Structure of solids, lattice energies

from the list of CHEMICAL CALCULATIONS

3. molar masses by gas density

4. gas laws, including the ideal gas law, Dalton's law and Graham's law

from the list of EQUATIONS & CONSTANTS

$PV = nRT$	the key relationship for success in managing gas laws problems
$\left(P + \dfrac{n^2a}{V^2}\right)(V - nb) = nRT$	Use this to address deviations from the general gas law. The van der Waals coefficients, a and b, have been determined for many gases and are found in standard handbooks. The "a" coefficient corrects for pressure that is not exerted by a gas because some attractive forces between molecules do exist for real gases. The "b" coefficient deducts from the available volume the amount of space actually occupied by the molecules. Larger molecules with stronger forces of attraction have relatively larger values for "a" and "b". Calculations using these formulas are cumbersome and unlikely to appear on the exam.
$P_A = P_{total} \cdot X_A$	Raoult's Law is somewhat scary way to say that the vapor pressure of component A in a solution is proportional to its concentration expressed as mole fraction. This is a rarely used concept.
$P_{total} = P_A + P_B + P_C + ...$	See discussion of Dalton's Law, as above.
$n = \dfrac{m}{M}$	How to calculate moles, mass or molar mass when two of those are known. You are in the wrong exam if you need this as reference information.
$K = {}^\circ C + 273$	Changing Celsius temperatures to kelvins.
$\dfrac{P_1V_1}{T_1} = \dfrac{P_2V_2}{T_2}$	Combining Charles's Law and Boyle's Law. Use this when you know or can figure out 5 of the 6 variables.
$D = \dfrac{m}{V}$	The definition of density, straight out of general science.
$u_{rms} = \sqrt{\dfrac{3kT}{m}} = \sqrt{\dfrac{3RT}{M}}$	root mean square speed information; obscure and cumbersome
KE per molecule $= \dfrac{1}{2}mv^2$	obscure and cumbersome
KE per mole $= \dfrac{3}{2}RT$	Use R in joules mol$^{-1}K^{-1}$
$\dfrac{r_1}{r_2} = \sqrt{\dfrac{M_2}{M_1}}$	See discussion above about Graham's Law. Extra cumbersome; follow the advice above.

from the list of RECOMMENDED EXPERIMENTS

3. determination of molar mass by vapor density

5. determination of molar volume of a gas

Multiple Choice Questions

Questions 1-5: Energy is added to a system containing a pure substance at a constant rate as shown in the warming curve below.

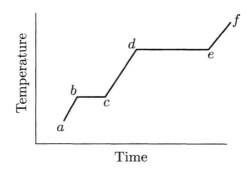

1. Which is most closely related to the length of the line segment $b - c$?

    (A) heat of fusion
    (B) heat of vaporization
    (C) specific heat capacity of the gas
    (D) specific heat capacity of the solid
    (E) melting point of the solid

2. Which region represents the greatest change in potential energy of the system?

    (A) $a - b$
    (B) $b - c$
    (C) $c - d$
    (D) $d - e$
    (E) $e - f$

3. Which accounts for the difference in length of the line segments $b - c$ and $d - e$?

    (A) The heat of fusion is less than the heat of vaporization.
    (B) The heat of fusion is greater than the heat of vaporization.
    (C) The solid has a greater specific heat capacity than the liquid.
    (D) The liquid has a greater specific heat capacity than the solid.
    (E) The heat of sublimation is equal to the sum of the heats of fusion and vaporization.

4. Which is the best description of the phase of the system as it changes from the conditions of point c to those of point d?

    (A) mixture of liquid and gas
    (B) mixture of liquid and solid
    (C) liquid with constant temperature
    (D) liquid with increasing temperature
    (E) liquid with decreasing temperature

5. Which region of the graph represents the condition where the rate of increase in the number of molecules in the gas phase per unit time is the greatest?

    (A) $a - b$
    (B) $b - c$
    (C) $c - d$
    (D) $d - e$
    (E) $e - f$

_____

6. Which could account for a change in the distance between molecules in a sample of gas?

    I.   increase in density at constant temperature
    II. increase in temperature at constant pressure
    III. increase in temperature at constant volume

    (A) I only
    (B) II only
    (C) I and II only
    (D) I and III only
    (E) I, II and III

7. The density of an unknown gas is found to be $3.00 \, \mathrm{g \, L^{-1}}$. Under the same conditions, the density of oxygen gas is found to be $2.00 \, \mathrm{g \, L^{-1}}$. The molar mass of the unknown gas is closest to

    (A) 14 g
    (B) 24 g
    (C) 36 g
    (D) 48 g
    (E) 72 g

8. Under certain conditions, methane gas, CH_4, diffuses at a rate of $12 \, \mathrm{cm \, sec^{-1}}$. Under the same conditions, an unknown gas diffuses at a rate of $8.0 \, \mathrm{cm \, sec^{-1}}$. The molar mass of the unknown gas is closest to

    (A) 6
    (B) 20
    (C) 24
    (D) 36
    (E) 72

9. In the van der Waals equation used to predict the behavior of real gases

$$(P + \frac{n^2 a}{V^2})(V - nb) = nRT$$

which characteristic is most closely related to the b term?

(A) forces of attraction between molecules
(B) effective volume of the molecules
(C) translational velocity of the molecules
(D) mass of the molecule
(E) force of collision between molecules

10. Which expression gives the volume in liters of 1.00 mole of an ideal gas at 20°C and 0.98 atm?

(A) $22.4 \times \dfrac{293}{273} \times \dfrac{1.00}{0.98}$

(B) $22.4 \times \dfrac{273}{293} \times \dfrac{1.00}{1.02}$

(C) $22.4 \times \dfrac{293}{273} \times \dfrac{0.98}{1.02}$

(D) $22.4 \times \dfrac{293}{273} \times \dfrac{0.98}{1.00}$

(E) $22.4 \times \dfrac{273}{293} \times \dfrac{1.00}{0.98}$

11. At certain conditions, the molar volume of a real gas may be less than the value predicted by the ideal gas law. Which property accounts for this deviation?

(A) Each gas molecule occupies an absolute volume.
(B) Forces of attraction exist between the gas molecules.
(C) Resonance bonds exist between the atoms in the molecules of the gas.
(D) The average velocity of the gas molecules is less than the value predicted by Graham's Law.
(E) The kinetic energy of the gas molecules is less than the value predicted by the $KE = \frac{1}{2}mv^2$.

12. Under the same conditions of pressure, sulfur dioxide liquifies at a much higher temperature than carbon dioxide. Which best accounts for this difference?

(A) Each sulfur dioxide molecule has a greater absolute volume than a carbon dioxide molecule.
(B) Stronger forces of attraction exist between sulfur dioxide molecules than between carbon dioxide molecules.
(C) S–O bonds illustrate resonance; C–O bonds do not.
(D) Each sulfur dioxide molecule has a greater molecular mass than a carbon dioxide molecule.
(E) At the same conditions of temperature and pressure, a sulfur dioxide molecule has greater density than a carbon dioxide molecule.

13. A sample of hydrogen gas is collected by displacement of water under ordinary conditions as shown in the diagram below. All of the following describe this sample of gas EXCEPT

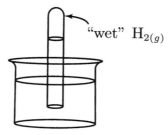

"wet" $H_{2(g)}$

(A) The volume occupied by $H_{2(g)}$ is greater than the volume occupied by $H_2O_{(g)}$.

(B) The pressure of $H_{2(g)}$ is greater than the pressure of $H_2O_{(g)}$.

(C) The temperature of $H_{2(g)}$ is the same as the temperature of $H_2O_{(g)}$.

(D) The number of molecules of $H_{2(g)}$ is greater than the number of molecules of $H_2O_{(g)}$.

(E) The mass of a molecule of $H_{2(g)}$ is less than the mass of a molecule of $H_2O_{(g)}$.

14. Which is the best description of water in the solid phase?

(A) atomic solid
(B) ionic solid
(C) network solid
(D) covalent solid
(E) molecular solid

15. Which property of hydrogen sulfide gas is least closely related to the polarity of its molecules?

(A) molar mass
(B) solubility in water
(C) critical temperature
(D) normal boiling point
(E) elasticity of molecular collisions

Questions 16-18: Consider three pistons each containing 2.50 g of the gas specified in 2.24 liters measured at 273 K. The pressure is not specified. Assume ideal behavior.

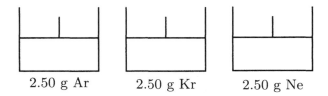

2.50 g Ar   2.50 g Kr   2.50 g Ne

16. Which is a correct comparison of the contents of the pistons?

    I.   The number of molecules in each piston is the same.
    II. The density of the contents of each piston is the same.
    III. The average velocity of the molecules in each piston is the same.

(A) II only
(B) III only
(C) I and II only
(D) II and III only
(E) I, II, and III

17. Which is a correct comparison of the contents of the systems in the pistons?

(A) All have the same pressure because they have the same mass and volume.

(B) All have the same pressure because they have the same average kinetic energy.

(C) All have the same pressure because noble gases deviate little from the ideal behavior.

(D) The pressure of the krypton is the greatest because the molar mass of krypton is the greatest.

(E) The pressure of the neon is the greatest because the number of neon molecules is the greatest.

18. If the pressure in each piston is adjusted to one atmosphere at constant temperature, which change will be observed?

(A) All volumes will change to the same lower value.

(B) All volumes will change to the same greater value.

(C) All volumes will increase such that the neon becomes greater than the other two.

(D) The volume of the neon will increase and the volumes of the other two will decrease.

(E) The volume of the neon and the argon will increase and the volume of the krypton will decrease.

19. Consider a sample of gas confined at constant temperature and volume in the closed system shown below. If more of this same gas is added at constant temperature, what effect is observed on pressure and average molecular velocity?

   (A) Both pressure and average molecular velocity increase.
   (B) Pressure decreases and average molecular velocity remains the same.
   (C) Pressure remains the same and average molecular velocity increases.
   (D) Pressure increases and average molecular velocity remains the same.
   (E) Pressure remains the same and average molecular velocity decreases.

20. When a sample of ideal gas is heated from 20°C to 40°C, the average kinetic energy of the system changes. Which factor describes this change?

   (A) $\dfrac{1}{2}$

   (B) $\dfrac{313}{293}$

   (C) $\sqrt{\dfrac{313}{293}}$

   (D) $\dfrac{293}{313}$

   (E) 2

21. Which occurs when a substance is converted from liquid to gas at the normal boiling point?

   I.   Potential energy of the system increases.
   II. The distance between molecules increases.
   III. The average kinetic energy of the molecules increases.

   (A) I only
   (B) II only
   (C) I and II only
   (D) II and III only
   (E) I, II, and III

22. The vapor pressure of diethyl ether, $C_2H_5OC_2H_5$, is greater than the vapor pressure of ethyl alcohol, C_2H_5OH at any temperature. Which best accounts for this difference?

    (A) The normal boiling point of ethyl alcohol is higher than the normal boiling point of diethyl ether.

    (B) Diethyl ether molecules have greater polarity than ethyl alcohol molecules.

    (C) The molar mass of diethyl ether is greater than the molar mass of ethyl alcohol.

    (D) In the liquid phase, the density of diethyl ether is greater than the density of ethyl alcohol.

    (E) The forces of attraction between molecules of ethyl alcohol are stronger than the forces of attraction between molecules of diethyl ether.

23. Solid carbon dioxide, dry ice, is observed to sublime at ordinary temperature and pressure. Which procedure is most likely to prevent the loss of solid carbon dioxide due to sublimation?

    (A) Submerge the solid carbon dioxide in a container of dense oil at ordinary conditions.
    (B) Change conditions to higher pressure and higher temperature.
    (C) Change conditions to higher pressure and lower temperature.
    (D) Change conditions to lower pressure and higher temperature.
    (E) Change conditions to lower pressure and lower temperature.

24. Consider the two sealed flasks with the same volume shown below, each containing 0.100 mol of the gas specified at STP.

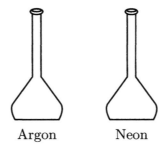

Argon   Neon

    Which gives the best comparison of the three properties of these systems specified below?

	(1) mass of the contents of the flask	(2) number of molecules	(3) average molecular velocity
(A)	Ne = Ar	Ne = Ar	Ne = Ar
(B)	Ne > Ar	Ne > Ar	Ne > Ar
(C)	Ne < Ar	Ne = Ar	Ne > Ar
(D)	Ne < Ar	Ne < Ar	Ne < Ar
(E)	Ne = Ar	Ne = Ar	Ne > Ar

25. Consider the sealed metal tank, shown below, containing an ideal gas at room temperature. When this system is cooled to 273 kelvins, some other properties change; some do not. Which gives a correct summary of changes in such properties?

  I. The density increases.
  II. The average molecular velocity decreases.
  III. The pressure of the system decreases.

(A) II only
(B) II and III only
(C) I and III only
(D) III only
(E) I, II,and III

Free-Response Questions

26. The equation of state $PV = n\text{RT}$ applies to an ideal gas. This relationship is sometimes called the General Gas Law or Ideal Gas Law.

    (A) Discuss two characteristics of gas molecules that account for nonideal behavior.

    (B) Why does cooling and/or compression cause most gases to change to the liquid phase?

    (C) What deviation from the predicted value for the PV product occurs at low temperature? Explain.

    (D) What deviation from the predicted value for the PV product occurs at high pressure? Explain.

27. Use principles of the Kinetic Molecular Theory to respond to the questions below.

    (A) In the space below, complete the sketches of two pistons to represent samples containing 1.60 g of the gas specified in 0.5 liters at 298 K. Each piston should be similar to the example shown.

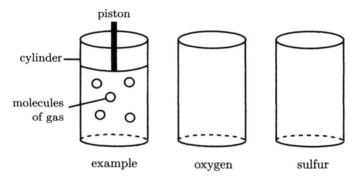

    Your sketches should show the correct comparison of relative volume and relative number of molecules in each sample.

    (B) Compare the average kinetic energy of each sample. Explain.
    (C) Compare the average molecular velocity for each sample. Explain.
    (D) Compare the pressure for each sample. Explain.

28. Phase diagrams are used to specify phase in a closed system as a function of pressure and temperature combinations.

(A) On the axes provided below, sketch a phase diagram for carbon dioxide. Include boundaries for each of the three phases. Clearly label the following on your diagram:

- each of the three phases
- triple point
- critical temperature

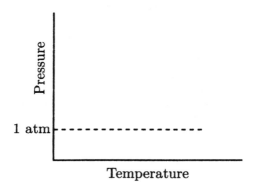

(B) Which features of the phase diagram provide the following information? Explain.

(1) Carbon dioxide does not have a normal freezing point or normal boiling point.

(2) The density of solid carbon dioxide is greater than the density of liquid carbon dioxide.

(C) A slush of dry ice and acetone can be used to provide a constant temperature bath for lab use in a properly ventilated area. Which point on the graph prepared in part (A) is most closely related to the expected temperature of the dry ice/acetone slush? Explain.

CHAPTER 4
SOLUTIONS

Solution concepts are major components of every Advanced Placement Chemistry examination. You should be able to address these concepts qualitatively by describing or explaining key ideas about solutes, solvents, solutions and the dissolving process. You will most certainly be required to characterize solutions quantitatively, including mass percent, density, molality and especially molarity.

TYPES OF SOLUTIONS AND FACTORS AFFECTING SOLUBILITY

A solution is made up of at least two components. The major component is called the solvent while the minor component is the solute. A solution can be made from any pair of the phases of matter. Most people think of a solution as a mixture with solid solute dispersed in a liquid solvent. However, any pair of phases is possible. For example, although it might not seem intuitively obvious, an alloy such as brass is a solution made of two metallic solids such as copper and zinc which retains many of their metallic characteristics.

	PARTS OF A SOLUTION
Solution	a homogeneous (single phase) mixture of two or more substances
Solvent	the component of a solution that defines its phase; the substance into which the solute is dissolved, melted, or dispersed
Solute	the substance that dissolves in a solvent to form a solution; usually dispersed as ions or molecules

Factors that affect solubility

Solvent-solute interaction. The solubility of a solute into a solvent depends on the characteristics of each component of the solution. Solutes and solvents with similar intermolecular forces tend to attract each other readily and form solutions easily. The stronger the intermolecular forces between a solvent and a solute, the more likely a solution is to form. The phrase that describes this is easy to remember: "like dissolves like". Thus, a polar solvent will dissolve a polar solute and a non-polar solute will be attracted easily into a non-polar solvent. A non-polar liquid will not be soluble in a polar liquid and instead will be immiscible, forming separate liquid layers that do not mix significantly in each other.

Temperature effect. Most solid solutes become more soluble in water as the temperature increases. You can dissolve more table sugar in hot water than in cold. This temperature effect is also true for many ionic substances in water. The opposite trend is true for gases forming a solution in water. The higher the temperature, the higher the kinetic energy of the gases in solution and the more likely they are to bubble out of solution.

Pressure effect. Gases are less likely to escape from a solution formed with a liquid solvent if high pressure is kept over the surface of the liquid mixture. The solubility of solids and liquids in solution is not affected by pressure. You can remember the effects on a gaseous solute in liquid solvent easily by thinking of your own behavior with a partially finished bottle of cola: you put the cap back on (to maintain pressure over the solution) and you place the bottle back into the refrigerator (to slow the gas molecules and keep them in solution).

Density and concentration

There are a number of ways of expressing concentration quantitatively but only the six defined below are commonly used in AP Chemistry. Note that density is not a measure of solution concentration. Any concentration measure that includes a volume is temperature-dependent (since volume of the solvent, especially, may vary with temperature). Any stated molarity, for example, should therefore include the temperature at which it was measured. Ratios that compare moles or mass are not temperature-dependent since those quantities do not change as the temperature changes.

Molarity includes a term expressing the volume of a solution. Because density is a ratio of mass of a solution to its volume, density is often used to calculate mass percent from molarity. With additional calculation, moles, mole fraction or molality can be determined.

Normality, formality, or other concentration measures are not tested.

QUANTITATIVE DESCRIPTIONS OF SOLUTIONS

$$\text{molarity } (M) = \frac{\text{mol solute}}{\text{(one) L solution}}$$

$$\text{molality } (m) = \frac{\text{mol solute}}{\text{(one) kg solvent}}$$

$$\text{mole fraction} = \frac{\text{mol solute}}{\text{mol solute} + \text{mol solvent}}$$

$$\text{pph (parts per hundred) or mass percent} = \frac{\text{g solute}}{100 \text{ g solution}}$$

$$\text{ppm (parts per million)} = 10^4 * \text{pph} \quad \text{used especially for dilute solutions}$$

$$\text{ppb (parts per billion)} = \text{ppm} * 10^3 \quad \text{used for extremely dilute solutions}$$

$$\text{density} = \frac{\text{g solution}}{\text{(one) mL solution}}$$

Colligative properties and Raoult's law (nonvolatile solutes); osmotic pressure

Colligative properties represent the characteristics of a solution that differ from those of the pure solvent due to the presence of solute particles. They include expansion of the range over which the solution remains liquid (boiling point elevation and freezing point elevation), lowering of vapor pressure, and change in osmotic pressure. These properties are determined by the number of solute particles (molecules or ions), not by the nature of those particles.

Colligative properties are described by a set of equations that indicate the difference in behavior between a solution and that of a sample of the pure solvent. The presence of particles of solute in the solvent interferes with the normal behavior of the pure solvent. The identity of the particles does not matter, just their presence. The quantity of particles present does matter, so a measure of concentration occurs in each equation. These equations describe the properties of dilute solutions best, often called "ideal behavior". At higher concentrations, the solutions no longer behave in an ideal fashion.

1. Boiling point elevation
$$\Delta T_b = iK_b m, \text{ where}$$

    i = number of moles of particles in solution (van't Hoff value)

    K_b = molal boiling point constant for a given solvent

    m = concentration in molality

2. Freezing point depression
$$\Delta T_f = iK_f m, \text{ where}$$

    i = number of moles of particles in solution (van't Hoff value)

    K_f = molal freezing point depression constant; particular to the solvent

    m = concentration in molality

3. Increase in osmotic pressure
$$\pi = MRT$$

    T in kelvins, $R = 0.0821$ (thermodynamic constant), where

    M = molarity

    R = the gas constant, 0.0821 L atm mol^{-1} K^{-1}

    T = temperature, kelvins

4. Decrease in vapor pressure

    where $P_{\text{vap, soln}} = X_{\text{solv}} \cdot P^{\circ}_{\text{vap, solv}}$

    $P_{\text{vap, soln}}$ = vapor pressure for solution

    X_{solv} = mole fraction of solvent

    $P^{\circ}_{\text{vap, solv}}$ = vapor pressure for pure solvent
    (assumes non-volatile solute)

Solubility

Solubility refers to the amount of solute that will dissolve in a given amount of solution or solvent at a specified temperature.

<u>Solubility Guidelines for ionic compounds</u>: this simplified set of five rules is very serviceable and allows accurate prediction of solubilities in most instances. If you need more precise information, you must consult a table of K_{sp} values. (More later in Chapter 7.)

All salts containing ..?.. (choose any ion from the lists below) **are soluble.**

The cation guideline:

1. Na^+, NH_4^+, K^+

The anion guidelines:

2. NO_3^-, $C_2H_3O_2^-$, ClO_3^-, ClO_4^-
3. Halides (Cl^-/Br^-/I^-)
   except when paired with an ion from the silver group: Ag^+, Hg_2^{2+}, Pb^{2+}
4. SO_4^{2-} except silver group (above) and Ba^{2+} and Sr^{2+}
5. Everything else is assumed to be insoluble.

Solubility product constant (K_{sp}) (See also Chapter 7.)

$$AgCl_{(s)} \rightleftharpoons Ag^+{}_{(aq)} + Cl^-{}_{(aq)}$$

$$K_{sp} = [Ag^+][Cl^-]$$

For a two ion salt (eg. AgCl), $K_{sp} = s^2$ where s is the molar solubility

three ion salt $= 4s^3$	(eg. $PbCl_2$)
four ion salt $= 27s^4$	(eg. $Al(OH)_3$)
five ion salt $= 108s^5$	(eg. $Ca_3(PO_4)_2$)

You should be able to convert easily from grams of solute per liter of solution to molarity of each dissolved species, thence to K_{sp}.

Three ion: $\qquad PbCl_2 \rightleftharpoons Pb^{2+} + 2Cl^- \qquad$
$$\begin{aligned} K_{sp} &= [Pb^{2+}][Cl^-]^2 \\ &= (s)(2s)^2 \\ &= 4s^3 \end{aligned}$$

Four ion: $\qquad Al(OH)_3 \rightleftharpoons Al^{3+} + 3OH^- \qquad$
$$\begin{aligned} K_{sp} &= [Al^{3+}][OH^-]^3 \\ &= (s)(3s)^3 \\ &= 27s^4 \end{aligned}$$

Five ion: $\qquad Ca_3(PO_4)_2 \rightleftharpoons 3Ca^{2+} + 2PO_4{}^{3-} \qquad$
$$\begin{aligned} K_{sp} &= [Ca^{3+}]^3[PO_4{}^{3-}]^2 \\ &= (3s)^3\,(2s)^2 \\ &= 108s^5 \end{aligned}$$

Further discussion of solubility equilibrium is found in Chapter 7.

from the TOPIC OUTLINE (website: apcentral.collegeboard.com)

II. States of Matter

C. Solutions

    1. Types of solutions and factors affecting solubility

    2. Methods of expressing concentration (The use of normalities is not tested.)

    3. Raoult's Law and colligative properties (nonvolatile solutes); osmosis

    4. Non-ideal behavior (qualitative aspects)

 from the list of CHEMICAL CALCULATIONS

3. Molar masses from gas density, freezing-point, and boiling-point measurements:

6. Mole fractions; molar and molal solutions

 from the list of EQUATIONS & CONSTANTS

$\Delta T_f = iK_f \times$ molality; $K_f =$ freezing point depression constant
       $i =$ van't Hoff factor

$\Delta T_b = iK_b \times$ molality (boiling point elevation)

   $\pi = MRT$

   $A = abc$

 from the list of RECOMMENDED EXPERIMENTS

4. Determination of molar mass by freezing-point depression

17. Colorimetric or spectrophoto metric analysis

Multiple Choice Questions

Questions 1-5: The set of lettered choices is a list of aqueous solutions of chloride compounds in various amounts. Select the one lettered choice that best fits each numbered description. A choice may be used once, more than once or not at all.

(A) 400 mL of 0.30 M NH_4Cl (molar mass: 53.5 g)

(B) 500 mL of 0.10 M NaCl (molar mass: 58.5 g)

(C) 200 mL of 0.20 M KCl (molar mass: 74.5 g)

(D) 100 mL of 0.20 M $MgCl_2$ (molar mass: 95.3 g)

(E) 200 mL of 0.10 M $FeCl_3$ (molar mass: 162 g)

1. contains the highest concentration of Cl^- ions

2. has the highest osmotic pressure

3. has the highest vapor pressure

4. has the greatest mass of solution

5. has the greatest mass of solute

---

6. What is the final concentration of Cl^- ion when 250 mL of 0.20 M $CaCl_2$ solution is mixed with 250 mL of 0.40 M KCl solution? (Assume additive volumes.)

   (A) 0.10 M

   (B) 0.20 M

   (C) 0.30 M

   (D) 0.40 M

   (E) 0.60 M

7. Which statement applies to a saturated aqueous solution of the highly soluble salt, sodium carbonate, Na_2CO_3, in contact with excess solid at constant temperature?

   I.   The system illustrates solubility equilibrium.
   II. The molarity of the solution is equal to its molality.
   III. The rate at which anions dissolve is less than the rate at which cations dissolve.

   (A) I only

   (B) III only

   (C) I and II only

   (D) I and III only

   (E) I, III, and III

8. A solution is prepared by dissolving 1.00 mol glycerol, a nonvolatile nonelectrolyte, in 9.00 mol water. The vapor pressure of water at 25°C is 23.8 mmHg. The vapor pressure of the solution in mmHg is

   (A) 0.147
   (B) 0.853
   (C) 2.38
   (D) 21.4
   (E) 23.8

9. A water solution of sucrose ($C_{12}H_{22}O_{11}$, molar mass: 342) is known to be 6.0 molal. What one additional characteristic of the solution could be used to determine its density?

   (A) mass
   (B) molarity
   (C) boiling point
   (D) freezing point
   (E) percent by mass

10. Which ratio gives the mole fraction of $HC_2H_3O_2$ when 60.0 g acetic acid ($HC_2H_3O_2$, molar mass: 60.0 g) is dissolved in 90.0 g water (H_2O, molar mass: 18.0 g)?

    (A) $\frac{1}{5}$
    (B) $\frac{1}{6}$
    (C) $\frac{2}{3}$
    (D) $\frac{2}{5}$
    (E) $\frac{11}{1}$

11. When used to prepare a standard solution of solid acid with specified molarity, which apparatus provides the greatest precision for measuring the specified quantity of solution to be prepared?

    (A) Dewar flask
    (B) volumetric flask
    (C) Erlenmeyer flask
    (D) analytical balance
    (E) centigram balance

12. What is the mass percent of ammonium dichromate in water solution if 25 g ammonium dichromate $((NH_4)_2Cr_2O_7$, molar mass: 252.1 g) is dissolved in 100. g water?

   (A)   1.8%
   (B)   20%
   (C)   25%
   (D)   80%
   (E)   100%

13. A standard solution of sodium hydroxide can be used in a titration experiment to determine the molar mass of a solid acid. A common mistake in such a titration experiment is the failure to rinse the buret with the standard solution after the final water rinse but before measurements of the volume of the standard solution are taken. This mistake accounts for which of the following results?

   I.   The reported volume of the standard solution used in the titration reaction is too small.

   II.   The reported volume of the solute used to dissolve the unknown acid is too small.

   III.   The reported number of moles of unknown acid used in the titration reaction is too large.

   (A)   I only
   (B)   II, and III only
   (C)   III only
   (D)   I and III only
   (E)   I, II, and III

14. The level of arsenic permitted in drinking water is 0.050 ppm (parts per million). Which of the following is another way to express that same concentration?

   (A)   0.050 mg As/milliliter H_2O
   (B)   0.050 mg As/liter H_2O
   (C)   0.050 g As/million liters H_2O
   (D)   0.050 mg As/million liters H_2O
   (E)   0.050 mg As/million grams H_2O

Questions 15 and 16: refer to the solubility curve for KCl in water as shown below.

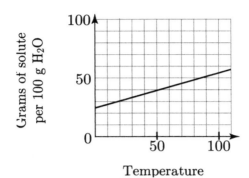

15. At what temperature is the concentration of a saturated solution of KCl (molar mass: 74.5 g) approximately 3 molal?

(A) 0°C
(B) 35°C
(C) 50°C
(D) 80°C
(E) 100°C

16. What is the mass percentage of water in a saturated solution of KCl at 80°C?

(A) 20%
(B) 33%
(C) 50%
(D) 67%
(E) 80%

Questions 17 and 18: refer to the solubility curve for KCl in water as shown above. A mixture containing 100 g H_2O and 40. g KCl is warmed to 60°C and stirred thoroughly. It is then cooled to 40°C with no immediate change in appearance.

17. The resulting solution is best described as

(A) colloidal
(B) isotonic
(C) unsaturated
(D) saturated
(E) supersaturated

18. When a tiny crystal of KCl is added to the cooled solution, a quantity of white crystalline solid forms. Which is the best description of the mass of the solid phase that forms in the system at 40° C?

(A) 0 g solid
(B) 2 g solid
(C) 10 g solid
(D) 20 g solid
(E) 55 g solid

19. Which accounts for the Tyndall effect in colloids?

    (A) scattering of light by particles of matter
    (B) absorption of light by particles of matter
    (C) absorption of light of specific wavelength in the visible range
    (D) absorption of light of specific wavelength in the ultraviolet range
    (E) alternating patterns of refraction and reflection of light by lattice particles

20. A hydrophobic colloid is most likely to be stabilized in water by the presence of

    (A) sodium ions, Na^+

    (B) benzene molecules, C_6H_6

    (C) hydrogen ions, H_3O^+

    (D) sucrose molecules, $C_{12}H_{22}O_{11}$

    (E) stearate ions, $C_{17}H_{35}COO^-$

21. Which applies to a 1.0 molar solution of potassium nitrate in water?

    I.   Adding water raises the freezing point.
    II. Adding water increases the vapor pressure of the solution.
    III. Adding water decreases the density of the solution.

    (A) I only
    (B) II and III only
    (C) I and III only
    (D) II and III only
    (E) I, II, and III

22. Compared to water, a 0.20 M solution of NaCl will have all of the following properties EXCEPT

    (A) greater density
    (B) lower vapor pressure
    (C) lower boiling point
    (D) lower freezing point
    (E) greater conductivity

23. In a spontaneous, exothermic dissolving process, which of these values has a negative sign?

    I. ΔG_{soln}

    II. ΔH_{soln}

    III. ΔT

    (A) I only
    (B) III only
    (C) I and II only
    (D) II and III only
    (E) I, II, and III

24. A saturated solution of KNO_3 in equilibrium with excess solute is prepared at 20°C. Which of the following describes the solution after the temperature of the system is increased to 40°C while still in contact with excess solute?

    I.   The molality of the solution increases.

    II. The solution remains saturated.

    III. The density of the solution increases.

    (A) II only
    (B) III only
    (C) I and III only
    (D) II and III only
    (E) I, II, and III

25. A dilute solution of NaCl is prepared at 20°C. Which of the following describes the solution after the temperature of the solution is increased to 40°C?

    I.   The vapor pressure of the solution increases.
    II. The number of ion pairs in solution increases.
    III. The difference between the freezing point and the boiling point of the solution increases.

    (A) I only
    (B) III only
    (C) I and II only
    (D) I and III only
    (E) I, II, and II

Free-Response Questions

26. Answer the following questions concerning propanone (acetone, C_3H_6O), a substance used to remove nail polish.

    (A) Draw a Lewis structure for propanone.

    (B) Is propanone expected to be soluble in water? Explain.

    (C) A solution is prepared by combining 30.0 mL propanone (density: 0.792 g mL^{-1}) with 50.0 mL 2-propanol (C_3H_7OH, density 0.785 g mL^{-1}). Assume that the volumes are additive.

    (1) Calculate the percent by mass of 2-propanol in the solution.
    (2) Calculate the percent by volume of propanone in the solution.
    (3) Calculate the mole fraction of propanone in the solution.

    (D) Compare the expected freezing point of solution described in part (B) to the freezing point of pure 2-propanol. Is the freezing point of the solution expected to be higher than, equal to, or lower than the freezing point of the pure solvent? Explain.

27. Answer each of the following questions related to the dissolving process.

    (A) Identify the two major energy changes that determine whether the dissolving of any solid in water is exothermic or endothermic. Define each energy change.

    (B) When an ionic solid dissolves in water, the sign for ΔS is positive. Explain.

    (C) Discuss the effect of each energy change defined in part (A) on the solubility of an ionic solid.

    (D) Discuss the role of free energy change, ΔG, in determining the solubility of a solute/solvent pair.

    (E) The dissolving process for ammonium nitrate, NH_4NO_3, in water is endothermic.

    (1) When 0.10 mol of NH_4NO_3 is added to 100 mL of water at 298 K, will the temperature of the resulting solution be higher than, lower than or the same as the initial temperature of the water? Explain.

    (2) How will this observation of temperature be different if the amount of NH_4NO_3 is doubled? Explain.

28. Answer the following questions related to the procedures for preparing solutions in the laboratory. Distilled water and ordinary laboratory equipment are available for use.

(A) Describe the measurements and procedures needed to prepare a 1.0 M (1.0 molar) solution of H_2SO_4 (molar mass 98 g) in water using 49 g of H_2SO_4. Concentrated sulfuric acid is nearly 100% H_2SO_4. It is available as a dense liquid with known specific gravity.

(B) Different procedures are used when the solution of H_2SO_4 is to be prepared at 1.0 m (1.0 molal) concentration? Explain.

(C) Which of the two solutions – 1.0 molar or 1.0 molal – has the greater percent by mass H_2SO_4? Explain.

(D) Compared to the dissolving of alcohol in water, what additional precautions should be taken when preparing a solution of sulfuric acid in water? Explain.

(E) The two major changes associated with the dissolving of any solid in water are crystal lattice energy and hydration energy. Which is larger for an endothermic dissolving process? Explain.

CHAPTER 5

STOICHIOMETRY

STOICHIOMETRY: CHEMICAL CALCULATIONS

Stoichiometry is the branch of chemistry that deals with quantitative measurements associated with chemical reactions. In practical terms for the AP chemistry student, the term refers to many of the quantitative problems encountered as listed below. You need to be able to use the mole concept for measuring quantities of chemical elements and compounds in chemical reactions.

<u>Chemical formulas</u>: formula mass and percentage composition; empirical and molecular formulas; hydrated salts

<u>Chemical elements and compounds</u>: Moles, molar mass, molar volume, the Avogadro number

<u>Chemical reactions</u>: mass, volume and moles of reactants and products; yield: theoretical, actual and percent yield; limiting reactant

Many students just call these "mole problems". **Dozens are for counting doughnuts, reams are for counting paper, moles are for counting atoms (and molecules).**

One way to look at quantitative problem solving is to identify the information provided, then identify the information to be determined. Effective management of stoichiometry problems is NOT based on memorizing a series of steps for each problem type. It is based on establishing a mental mindset for yourself that produces a reliable problem-solving strategy.

One mole of any substance is one gram-formula-weight of that substance
  18 grams of water, H_2O
  98 grams of sulfuric acid, H_2SO_4
  342 grams of sucrose, $C_{12}H_{22}O_{11}$

Any one molecule of sulfuric acid, H_2SO_4, weighs about 5 times as much as a molecule of water, H_2O.

One mole is also the Avogadro number of particles of the substance identified.

  6.02×10^{23} molecules of water, H_2O
  6.02×10^{23} molecules of sulfuric acid, H_2SO_4
  6.02×10^{23} molecules of sucrose, $C_{12}H_{22}O_{11}$

We use this information in factors that look like this

$$\frac{18 \text{ g } H_2O}{1 \text{ mol } H_2O} \qquad \frac{1 \text{ mol } C_{12}H_{22}O_{11}}{342 \text{ g } C_{12}H_{22}O_{11}}$$

$$\frac{6.02 \times 10^{23} \text{ molecules } H_2SO_4}{1 \text{ mol } H_2SO_4}$$

$$\frac{6.02 \times 10^{23} \text{ atoms S}}{1 \text{ mol S}}$$

$$\frac{6.02 \times 10^{23} \text{ ions } Na^+}{1 \text{ mol } Na^+ \text{ ions}}$$

We use moles to help us count atoms by weighing. We can count by weighing because all molecules of a given substance have the same mass (weight).

Chemical Formulas

When dealing with chemical formula-oriented problems, a situation maybe presented where a chemical formula is known and some other information is to be determined. A chemical formula tells the relationship between numbers and types of atoms in a substance, perhaps as simple He (helium gas) or more complex such as $(NH_4)_4Fe(CN)_6 \cdot 3H_2O$, (ammonium ferrocyanide trihydrate).

When the formula is given, the problem assigned is likely to be to count the atoms of each component, then determine the mass percent of one or more components.

You probably established the mindset for this kind of problem a long time ago:
- get the mass of each part
- add up all the parts to get a total mass
- then get a percent for each.

In the opposite case, information on mass or percent by mass is given. You are assigned to calculate the ratio of atoms – that is the empirical formula (*empirical* means based on observation or experiment; in the context of chemistry problem-solving, an empirical formula is the simplest whole number ratio of atoms in the formula for a chemical compound). Alternatively, you may be asked to use a set of information to determine a chemical formula. Be sure to make full use of any information given. For example, if the information supplied says a hydrated binary salt of osmium is 54.2% by mass osmium and 15.4% water, what is the formula of the hydrated salt? Additional information that can be inferred is that the percent by mass of chlorine is 30.4. From there it is small step in logic to determine how to solve this problem.

You need to determine the ratio of atoms. First, you can look up how much one mole of atoms weighs. Then you can easily find out how many atoms of each element is contained in a known sample.

mass of atoms → number of moles of atoms → ratio of number of moles of atoms

In the case of this problem, the most convenient sample is 100. grams, from which you can immediately claim the composition as 54.2 g Os, 30.4 g Cl, and 15.4 g H_2O. Now, you can look up the atomic mass of each element (and the molar mass of H_2O - in case you forgot that it's 18 g/mol). Next, determine the number of moles of atoms of each; then you get a fractional ratio of moles of atoms of each which you can simplify to smallest whole number since you already passed 7th grade math.

$$54.2 \text{ g Os} \times \frac{1 \text{ mol Os}}{190 \text{ g Os}} = 0.285 \text{ mol Os}$$

$$30.4 \text{ g Cl} \times \frac{1 \text{ mol Cl}}{35.5 \text{ g Cl}} = 0.856 \text{ mol Cl}$$

$$15.4 \text{ g H}_2\text{O} \times \frac{1 \text{ mol H}_2\text{O}}{18 \text{ g H}_2\text{O}} = 0.856 \text{ mol H}_2\text{O}$$

Ratio: $Os_{0.285}$ $Cl_{0.856}$ $(H_2O)_{0.856}$

$Os_{\frac{0.285}{0.285}}$ $Cl_{\frac{0.856}{0.285}}$ $(H_2O)_{\frac{0.856}{0.285}}$

$$OsCl_3 \cdot 3H_2O$$

When more about moles is needed

Pushing a little further ahead, the information provided could be formula, mass, moles, volume of a gas at specified conditions, actual numbers of molecules, ions or atoms or even mass per unit volume (density). The assigned problem could be to determine one or more of the components of information not given.

The starting point for any of this is the molar mass using atomic masses from the Periodic Table. Start from anywhere, get moles, then go anywhere else. Figure 5.1 will help organize your strategies for solving "mole problems".

Figure 5.1

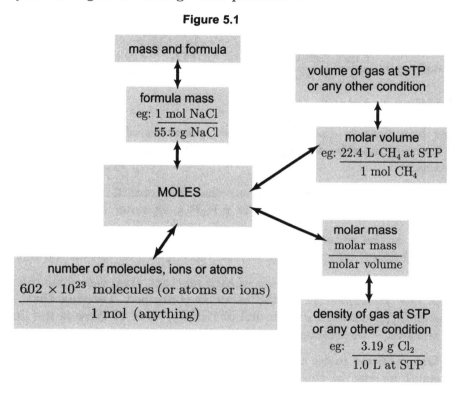

NOW FOR SOME ACTUAL CHEMICAL REACTIONS

Substances react one atom or even one mole of atoms at a time. In order to deal with quantities of reactants and products, we need a way to count the atoms. Because all atoms or molecules of a given substance have the same mass, it is convenient to count atoms or molecules by weighing.

Quantitative problems dealing with chemical reactions typically present a reaction with its corresponding chemical equation and one or more quantities of reactants. You could also be assigned to provide a desired quantity of product. Your task is to determine an unknown quantity such as quantity of product expected to be produced, and asked to determine the quantity of reactant to be supplied. In its simplest form, a question might ask you to determine what mass of silver phosphate is produced when 2.35 g $AgNO_{3(s)}$ is added to excess $H_3PO_{4(aq)}$. ("Excess" means that there is more than enough H_3PO_4 to react with all the $AgNO_3$. It also means that you don't need to think about a "limiting reactant" - see below.)

You need to write the balanced chemical equation for this reaction (looks like double replacement)

$$3AgNO_3 + H_3PO_4 \rightarrow Ag_3PO_4 + 3HNO_3$$

and remind yourself that the equation says

three moles $AgNO_3$ of reacts with one mole of H_3PO_4 to form one mole of Ag_3PO_4 and three moles of HNO_3

Once you get moles of $AgNO_3$, you can get moles, then mass, of anything else in this reaction.

$$2.35 \text{ g AgNO}_3 \times \frac{1 \text{ mol AgNO}_3}{170 \text{ g AgNO}_3} \times \frac{1 \text{ mol Ag}_3PO_4}{3 \text{ mol AgNO}_3} \times \frac{419 \text{ g Ag}_3PO_4}{1 \text{ mol Ag}_3PO_4} = 1.93 \text{ g Ag}_3PO_4$$

The limiting reactant

A further complication emerges when two quantities of reactants are given, then the limiting reactant must be determined and that value used to determine other quantities in the reaction.

In terms of the problem above, let's say that the conditions became an experiment in which 2.35 g $AgNO_3(s)$ is added to 1.58 g H_3PO_4 in water solution.

This is the same problem as above except that there are now two possibilities: either you use all the $AgNO_3$ or you use all the H_3PO_4.

You need to determine moles of each reactant, then determine the moles of one consumed if all the other is consumed, knowing that 3 mol $AgNO_3$ is needed for each mol of H_3PO_4. One of these is the limiting reactant – the reactant that is totally consumed.

Once you get moles of moles of limiting reactant, you can get moles of anything else in this reaction including moles of the excess reactant consumed and moles left unreacted. Once you have moles of these substances, you can determine any mass that is needed.

... and how much product you actually get – YIELD

There are two kinds of yield – how much that you predict can be produced based on moles of reactants and how much is actually produced that you can collect and weigh (count the atoms and molecules) after the reaction is completed. The predicted quantity is the amount determined as above, where we assumed full conversion to product. In actual practice, side reactions and other confounding events occur to decrease the amount of product. The amount actually produced is the actual yield. Percent yield is the fractional expression of mass actually produced as a fraction of the amount that could have been produced. We express that fraction as grams (or moles) actually produced per 100 grams (or moles) theoretically produced – parts per hundred or per cent.

Mindset: same as above for theoretical moles of product. Then weigh (count the atoms and molecules) the product that is actually produced. Convert this to a fraction with 100 grams of theoretical yield as the denominator.

from the TOPIC OUTLINE (website: apcentral.collegeboard.com)

III. Reactions

B. Stoichiometry

    1. Ionic and molecular species present in chemical systems; net ionic equations

    2. Balancing of equations including those for redox reactions

    3. Mass and volume relations with emphasis on the mole concept, including empirical formulas and limiting reactants

from the list of CHEMICAL CALCULATIONS

1. Percentage composition

2. Empirical and molecular formulas from experimental data

3. Molar masses from gas density, freezing-point, and boiling point measurements

5. Stoichiometric relations using the concept of the mole; titration calculations

Stoichiometry – a range of quantitative problems through stoichiometry of mixtures and limiting/excess reactants

from the list of EQUATIONS & CONSTANTS

Believe it or not, this table actually identifies n as [number of] moles, m as mass and M as molar mass (whenever it doesn't mean molarity).
Avogadro's number $= 6.022 \times 10^{23}$ [molecules] mol^{-1} is also provided.

The relation $n = m/M$ is also given. If you need that as a reference, you are in the wrong exam.

from the list of RECOMMENDED EXPERIMENTS

9. Determination of mass and mole relationship in a chemical reaction

16. Analytical gravimetric determination

Nearly every laboratory activity calls for use of stoichiometry – counting the atoms and molecules – that are being used or produced. Whenever the directions include use of weighed quantities, these are actually quantities with a known number of moles of molecules, ions or atoms.

Multiple Choice Questions

1. Which of these alkaline earth metal oxides has the greatest percent by mass of oxygen?

    (A) barium oxide
    (B) beryllium oxide
    (C) calcium oxide
    (D) magnesium oxide
    (E) strontium oxide

2. Which expression gives percent by mass of carbon in oxalic acid, $H_2C_2O_4 \cdot 2H_2O$?

    (A) $\dfrac{2}{14} \times 100$

    (B) $\dfrac{12}{90} \times 100$

    (C) $\dfrac{24}{66} \times 100$

    (D) $\dfrac{24}{90} \times 100$

    (E) $\dfrac{24}{126} \times 100$

3. Which oxides of manganese, Mn, have percent by mass of manganese that is greater than 50%?

    | | I. | MnO |
    | | II. | MnO_2 |
    | | III. | Mn_2O_3 |

    (A) II only
    (B) III only
    (C) I and III only
    (D) II and III only
    (E) I, II, and III

4. Which describes the resulting system when 0.40 moles of $Na_2CO_{3(s)}$ is added to 0.50 liters of 0.60 molar $CuCl_2$ solution?

    (A) A blue precipitate forms; excess $CO_3{}^{2-}$ is found in solution.
    (B) A blue precipitate forms; excess Cu^{2+} is found in solution.
    (C) A blue precipitate forms; no excess reactants are found in solution.
    (D) A nearly colorless homogeneous system forms; excess $CO_3{}^{2-}$ is found in solution.
    (E) A nearly colorless homogeneous system forms; excess Cu^{2+} is found in solution.

5. Which pair of samples contains the same number of oxygen atoms in each compound?

(A) 0.10 mol Al_2O_3 and 0.50 mol BaO

(B) 0.20 mol Cl_2O and 0.10 mol HClO

(C) 0.20 mol SnO and 0.20 mol SnO_2

(D) 0.10 mol Na_2O and 0.10 mol Na_2SO_4

(E) 0.20 mol $Ca(OH)_2$ and 0.10 mol $H_2C_2O_4$

6. Consider the reaction

$$2Na_3PO_{4(aq)} + 3ZnCl_{2(aq)} \rightarrow Zn_3(PO_4)_{2(s)} + 6\,NaCl_{(aq)}$$

A precipitate is formed when 0.20 moles of sodium phosphate, Na_3PO_4, is mixed with 0.80 moles of zinc chloride, $ZnCl_2$, in water solution. Which lists the ions in water solution after the reaction occurs, in order of increasing concentration?

(A) Na^+, Cl^-, Zn^{2+}, PO_4^{3-}

(B) Zn^{2+}, PO_4^{3-}, Na^+, Cl^-

(C) PO_4^{3-}, Zn^{2+}, Na^+, Cl^-

(D) PO_4^{3-}, Zn^{2+}, Cl^-, Na^+

(E) Zn^{2+}, PO_4^{3-}, Cl^-, Na^+

7. Consider the reaction

$$2ZnS_{(s)} + 3O_{2(g)} \rightarrow 2ZnO_{(s)} + 2SO_{2(g)}$$

Which value is closest to the mass of $ZnO_{(s)}$ produced when 50.0 g $ZnS_{(s)}$ is heated in an open vessel until no further weight loss is observed?

(molar masses: O_2 − 32 g, SO_2 − 64 g, ZnO − 81 g, ZnS − 97 g)

(A) 25 grams

(B) 40 grams

(C) 50 grams

(D) 60 grams

(E) 75 grams

8. Consider the reaction

$$2Al_{(s)} + 3Cl_{2(g)} \rightarrow 2AlCl_{3(s)}$$

Which expression gives the volume of Cl_2 consumed, measured at 1 atm and 273 K, when 25.0 g Al reacts completely with Cl_2 according to the above equation?

(A) $25.0 \times \dfrac{3}{2} \times \dfrac{22.4}{2}$

(B) $\dfrac{25.0}{22.4} \times \dfrac{3}{2} \times \dfrac{22.4}{2}$

(C) $25.0 \times \dfrac{27}{1} \times \dfrac{3}{2} \times \dfrac{22.4}{1}$

(D) $25.0 \times \dfrac{1}{27} \times \dfrac{3}{2} \times \dfrac{22.4}{1}$

(E) $25.0 \times \dfrac{1}{27} \times \dfrac{2}{3} \times \dfrac{22.4}{1}$

9. Which sample contains the greatest number of nitrogen atoms? (All measurements taken at 1 atm and 273 K.)

(A) 0.20 mol $N_2O_{4(g)}$

(B) 0.40 mol $N_{2(g)}$

(C) 40. L $NO_{2(g)}$

(D) 40. g $NH_{3(g)}$

(E) 80. g $N_2O_{4(g)}$

10.
$$C_3H_{8(g)} + 5O_{2(g)} \rightarrow 3CO_{2(g)} + 4H_2O_{(g)}$$

Propane gas, $C_3H_{8(g)}$, burns according to the equation above. A mixture containing 0.030 moles of $C_3H_{8(g)}$ and 0.200 moles of $O_{2(g)}$ is placed in a rigid container and its pressure measured. The mixture is ignited. Which describes the contents of the container after maximum reaction has occurred and the system returned to its original temperature?

(A) 0.050 mol $O_{2(g)}$ remains unreacted and the pressure has increased.

(B) 0.050 mol $O_{2(g)}$ remains unreacted and the pressure has decreased.

(C) 0.170 mol $O_{2(g)}$ remains unreacted and the pressure has decreased.

(D) 0.020 mol $C_3H_{8(g)}$ remains unreacted and the pressure has decreased.

(E) 0.020 mol $C_3H_{8(g)}$ remains unreacted and the pressure has increased.

11. When the equation for the reaction below is balanced using smallest whole numbers, which gives a correct description of the information in the equation?

$$..?..\ Sc(NO_3)_3 + ..?..\ NaOH \rightarrow ..?..\ Sc(OH)_3 + ..?..\ NaNO_3$$

   I.   The number of ions represented is 20.
   II. The number of atoms represented is 22.
   III. The sum of the coefficients is 8.

(A) II only
(B) III only
(C) I and II only
(D) II and III only
(E) I, II, and III

12. An unidentified compound is reported to contain 77.5% manganese and 22.5% oxygen by mass. Which set of values when substituted for x and y gives the best representation of the empirical formula for the unidentified compound?

$$Mn_xO_y$$

(A) $\dfrac{77.5}{54.9}$ and $\dfrac{16.0}{22.5}$

(B) $\dfrac{77.5}{54.9}$ and $\dfrac{22.5}{16.0}$

(C) $\dfrac{54.9}{22.5}$ and $\dfrac{16.0}{77.5}$

(D) $\dfrac{54.9}{77.5}$ and $\dfrac{16.0}{22.5}$

(E) $\dfrac{54.9}{22.5}$ and $\dfrac{16.0}{77.5}$

13. What minimum volume of 0.200 M Na_2CO_3 is needed to precipitate all the Sr^{2+} from 25.0 mL of 0.100 M $Sr(NO_3)_2$?

(A)   6.25 mL
(B)   12.5 mL
(C)   25 mL
(D)   50 mL
(E)   100 mL

14. $$Cu_{(s)} + 4HNO_{3(aq,conc)} \rightarrow 2NO_{2(g)} + Cu(NO_3)_{2(aq)} + 2H_2O$$

What volume of $NO_{2(g)}$ measured at 1 atm and 273 K can be produced by the reaction of 0.750 mol copper with excess concentrated nitric acid according to the equation above?

(A) 11.2 liters
(B) 22.4 liters
(C) 33.6 liters
(D) 44.8 liters
(E) 67.2 liters

15. $$Al^{3+} + 3e^- \rightarrow Al^0$$

In the half-reaction shown above, the term $3e^-$ represents

(A) $3 \times 96,500$ electrons

(B) $3 \times 6.02 \times 10^{23}$ electrons

(C) $3 \times 6.02 \times 10^{23}$ coulombs

(D) $\dfrac{3}{27} \times 96,500$ electrons

(E) $\dfrac{3}{27} \times 6.02 \times 10^{23}$ electrons

16. Consider the combustion of ethane as shown in the equation below.

$$2C_2H_{6(g)} + 7O_{2(g)} \rightarrow 4CO_{2(g)} + 6H_2O_{(g)}$$

What quantity of reactant remains after ignition of a mixture that contains 0.60 moles of C_2H_6 mixed with 2.50 moles of O_2? (Assume maximum reaction according to the equation above.)

(A) 0.20 mol O_2
(B) 0.40 mol O_2
(C) 1.90 mol O_2
(D) 0.20 mol C_2H_6
(E) 0.30 mol C_2H_6

Questions 17-19: A mixture is prepared by adding 100. mL of 0.10 M Na_2CrO_4 to 100. mL of 0.10 M $AgNO_3$. A precipitate forms in this mixture. The precipitate is separated from the mixture by filtration.

17. What is the concentration of Na^+ in the reaction mixture after filtration?

    (A) 0.050 M
    (B) 0.10 M
    (C) 0.15 M
    (D) 0.20 M
    (E) 0.40 M

18. What quantity of solid product is produced?

    (A) 0.0025 mol
    (B) 0.0050 mol
    (C) 0.010 mol
    (D) 0.015 mol
    (E) 0.020 mol

19. Which describes the changes in concentration of the spectator ions, Na^+ and NO_3^- in the reaction mixture as the reaction occurs in the beaker containing $AgNO_{3(aq)}$?

	$[Na^+]$	$[NO_3^-]$
(A)	increases	remains the same
(B)	remains the same	decreases
(C)	remains the same	remains the same
(D)	increases	decreases
(E)	decreases	decreases

---

20. $$2Al_{(s)} + 3S_{(s)} \rightarrow Al_2S_{3(s)}$$

    What mass of Al_2S_3 is produced when 1.50 moles of aluminum reacts with excess sulfur according to the equation above?

    (A) 40.5 g
    (B) 48.0 g
    (C) 61.5 g
    (D) 75.0 g
    (E) 113 g

21. When 0.60 mol ZnS was roasted in pure oxygen, 0.40 mol SO_2 was collected. Which best describes the contents of the solid phase remaining in the crucible?

    (A) no excess ZnS; 0.60 mol ZnO
    (B) no excess ZnS; 0.30 mol ZnO
    (C) 0.20 mol excess ZnS; 0.30 mol ZnO
    (D) 0.20 mol excess ZnS; 0.40 mol ZnO
    (E) 0.30 mol excess ZnS; 0.40 mol ZnO

22. The mass of element X found in 1.0 mole each of four different compounds is 28 g, 42 g, 56 g and 84 g, respectively. Which of the following is a possible atomic mass for element X?

   (A) 14
   (B) 28
   (C) 35
   (D) 42
   (E) 49

23. Which value is closest to the volume of $O_{2(g)}$ measured at STP that could be produced when 0.20 mol $KClO_{3(s)}$ is heated according to the equation below?

$$2KClO_{3(s)} \rightarrow 2KCl_{(s)} + 3O_{2(g)}$$

   (A)   4.5 L
   (B)   6.7 L
   (C)   7.5 L
   (D) 15. L
   (E) 34. L

24. How many moles of $KCl_{(s)}$ should be added to 0.500 liters of 0.20 M $CrCl_3$ solution to increase the chloride concentration to 1.00 M? (Assume no change in volume.)

   (A) 0.20
   (B) 0.40
   (C) 0.50
   (D) 0.60
   (E) 0.80

25. Epsom salt, $MgSO_4 \cdot 7H_2O$, (molar mass: 246 g) can be dehydrated by heating in an open crucible. Which value is closest to the fraction of the mass of salt in the crucible lost when the crucible is heated to constant weight?

   (A) $\dfrac{1}{8}$

   (B) $\dfrac{1}{7}$

   (C) $\dfrac{1}{4}$

   (D) $\dfrac{1}{3}$

   (E) $\dfrac{1}{2}$

Free-Response Questions

26. An unknown hydrocarbon, C_xH_y, exists as two isomers at 298 K; one is a gas, the other is a liquid.

    (A) When 3.90 grams of this hydrocarbon is placed in a 3.00 liter container and heated to 800. K, no liquid remains and the pressure becomes 1.58 atm. What is the molar mass of this hydrocarbon?

    (B) When the hydrocarbon is burned in oxygen, the only products formed are water and carbon dioxide. The mixture of products is 76.5% by mass $CO_{2(g)}$. What is the empirical formula of this hydrocarbon?

    (C) What is the molecular formula of this hydrocarbon? What is the IUPAC name of this hydrocarbon? Write the balanced equation for the complete combustion of this hydrocarbon.

27. A common rocket propellant is a mixture that includes liquid hydrazine, N_2H_4, as a fuel and liquid hydrogen peroxide, H_2O_2, as oxidizer. The products formed are nitrogen, $N_{2(g)}$, and water $H_2O_{(g)}$.

    (A) Write a balanced chemical equation for the reaction of N_2H_4 with H_2O_2.

    (B) What volume of $N_{2(g)}$ measured at 0.975 atm and 298 K is produced when 100. grams of $H_2O_{2(l)}$ is consumed?

    (C) Some standard enthalpies of formation ΔH_f°, at 298 K are given below:

$H_2O_{(g)}$	-241.6 kJ mol^{-1}
$N_2H_{4(l)}$	50.2 kJ mol^{-1}
$H_2O_{2(l)}$	-192.3 kJ mol^{-1}

    Using these values, calculate the standard enthalpy of reaction, ΔH°, for the reaction written in part (A).

    (D) If all of the energy produced by the reaction in part (B) is transferred to 5.00 kg of water at 20.0°C, what is the final temperature of the water? (The specific heat capacity of water is 4.18 J g^{-1} °C^{-1}.)

28. When iron is treated with excess steam and heated intensely, it is converted to Fe_3O_4, the laboratory version of the mineral, magnetite. Hydrogen gas is the other product of this reaction.

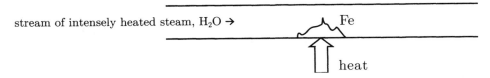

stream of intensely heated steam, H_2O →   Fe

heat

(A) Write the balanced equation for this reaction.

(B) Calculate the oxidation number of iron in Fe_3O_4.

(C) A sample of 25.0 g iron is heated intensely under a stream of hot steam. After this treatment, the increase in mass of the solid residue is 3.75 g.

  (1) Determine the percent by mass of Fe_3O_4 in the solid residue.

  (2) Calculate the number of moles of iron that remains unchanged.

(D) In the reaction described in part (C), what volume of hydrogen gas, measured at STP, was produced?

CHAPTER 6
CHEMICAL KINETICS

The study of reaction rates and mechanisms, also known as kinetics, is a key topic of the AP Chemistry program. Kinetics from the perspective of the differential rate law has had substantial exposure in many AP Exams over the years. Coverage of integrated rate law and the Arrhenius equation was recently restored to the AP Chemistry curriculum. You should be prepared to address reaction kinetics on both sections of the exam.

MOLECULES (AND IONS) IN COLLISION

You should be able to visualize the fundamental ideas of reaction kinetics via the **collision model**, an extension of the Kinetic Molecular Theory (KMT) that helps explain chemical change. The collision model describes chemical reactions as the summary effect of many, many collisions between potential reactant particles. Most collisions are ineffective and do not lead to a chemical reaction. Those that are effective "react" - that is, atoms get rearranged to produce a different substance.

Collisions

In order to be part of an **effective collision**, the reacting particles must be traveling fast enough – with enough kinetic energy – so that their collision transfers sufficient energy to break existing bonds. Furthermore, to be effective, this collision must occur with sufficiently favorable orientation so that new chemical bonds can form. Thus an effective collision must satisfy both a **kinetic energy** requirement and an **orientation** (or **steric**) requirement. You should be able to explain why an increase in temperature leads to more collisions, as well as more effective collisions. You should be able to explain why increases in concentration and surface area allow for more collisions. **Catalysts** change the rate of a reaction by diminishing the activation energy of a reaction. You should be able to make a sketch that shows why a surface catalyst diminishes the activation energy by trapping one reactant in a favorable orientation. A summary of the factors that affect rate of reaction is found in Figure 6.1. You should be able to explain why the presence of a catalyst accounts for more effective collisions. The **Arrhenius equation** incorporates both of these ideas.

$$\ln\left(\frac{k_2}{k_1}\right) = \frac{E_a}{R}\left(\frac{1}{T_1} - \frac{1}{T_2}\right) \qquad R = 8.134 \ \text{J mol}^{-1} \ \text{K}^{-1}$$

It is usually presented in a line-slope format that compares frequency of effective collisions at two different temperatures. Calculating the slope of the line thus drawn allows you to determine the **activation energy** of the reaction.

More and more effective collisions. **Reaction rate** is usually measured in mols, mL or mmHg per unit time. Reaction rate increases when temperature or concentration increases or when a catalyst is present. Using principles from the KMT and the collision theory, you should be able to explain why such changes occur. Figure 6.2 provides a good summary.

Figure 6.1 Factors affecting the rate of a reaction

- Temperature - at higher temperature, molecules collide with greater transfer of energy

- Concentration, including pressure for gas phase systems - at higher concentrations, the number of collisions increases

- Phase/surface/homogeneity of the system - provides for more favorable collision geometry

- Presence of a catalyst - catalyst provides a more favorable activated complex with lower requirement for energy of activation.

Reaction mechanism and reaction order

A **reaction mechanism** is a step-by-step description of the series of events, usually collisions, that lead to a reaction. The slowest step in that series determines the overall rate of the reaction and is known as the **rate determining step**. The **molecularity** of the reaction describes the number of reacting particles in the collision and can be uni-, bi- or (rarely) ter-molecular.

The **reaction order** describes the effect on the overall rate of a change in the concentration of a specific reactant. **First order,** where the rate change is directly proportional to the change in reactant concentration, is most common. **Second order** reactants change rate proportional to the square of the change in reactant concentration. Changes in concentration for **Zero** order reactants have no effect on the overall rate of the reaction. A **rate law** shows the overall rate of reaction as a function of the **rate constant** and the concentration of each reactant raised to the appropriate order. Figure 6.2 provides a good comparison of reaction orders.

Figure 6.2 Chemical Kinetics: comparing reaction order

	ORDER		
	Zero	**First**	**Second**
Rate law	$\text{Rate} = k[A]^0$	$\text{Rate} = k[A]^1$	$\text{Rate} = k[A]^2$
Integrated rate law	$[A] = -kt + [A]_0$	$\ln[A] = -kt + \ln[A]_0$	$\dfrac{1}{[A]} = kt + \dfrac{1}{[A]_0}$
Plot needed to give a straight line	$[A]$ versus t	$\ln[A]$ versus t	$\dfrac{1}{[A]}$ versus t
Relationship of rate constant to the slope of straight line	$\text{Slope} = -k$	$\text{Slope} = -k$	$\text{Slope} = k$
Half-life	$t_{\frac{1}{2}} = \dfrac{[A]_0}{2k}$	$t_{\frac{1}{2}} = \dfrac{0.693}{k}$	$t_{\frac{1}{2}} = \dfrac{1}{k[A]_0}$

The Maxwell-Boltzmann energy distribution graph in Figure 6.3 shows the distribution of kinetic energies for the same sample of gas molecules at different temperatures. Note that the curve at higher temperature has a greater fraction of molecules that possess the minimum energy required to start the reaction, E_a.

The graph of concentration of reactants as in Figure 6.4 (starting high and diminishing over time) and the products (starting low and increasing over time) shows that a <u>chemical equilibrium</u> is reached when both curves reach a horizontal line.

Figure 6.3 Energy distribution

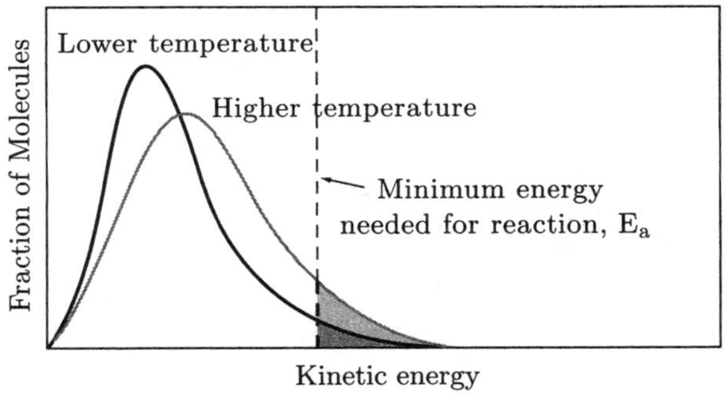

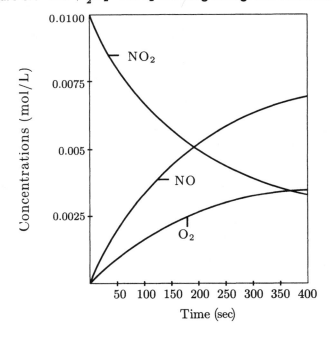

Figure 6.4 $NO + \frac{1}{2}O_2 \rightarrow NO_2$ Tracking changes in concentration

VOCABULARY: REACTION KINETICS

Reaction rates

Kinetics the study of reaction rates and mechanisms

Reaction Rate number of molecular events occurring per unit of time ($\Delta C/\Delta t$)

For the given reaction: $aA + bB \rightarrow dD + eE$

Overall Reaction rate: $\dfrac{-1}{a}\dfrac{\Delta[A]}{\Delta t} = \dfrac{-1}{b}\dfrac{\Delta[B]}{\Delta t} = \dfrac{1}{d}\dfrac{\Delta[D]}{\Delta t} = \dfrac{1}{e}\dfrac{\Delta[E]}{\Delta t}$

Relative rate of consumption: $\dfrac{-\Delta[A]}{\Delta t}$

Half-life time required for the concentration of a reactant to decrease by one-half of its original value

Catalyst substance that increases the rate of a reaction by providing a mechanism with lower activation energy for the reaction

Reaction mechanisms

Reaction mechanism the stepwise description of the processes by which reactant species change into products

Elementary Reaction (or process)– one step within a reaction mechanism

- **Unimolecular**– one molecule breaks into two pieces or undergoes a rearrangement to form a new isomer

- **Bimolecular** – two molecules collide and combine or transfer atoms

- **Termolecular** – three molecules collide and combine or transfer atoms

Rate-determining step the one elementary step in the reaction mechanism that is significantly slower than others. This step limits or determines the overall rate of the reaction.

Activation Energy (E_a) energy barrier that must be overcome in order for reactants to convert to products; potential energy acquired during the formation of the activated complex from the kinetic energy of the colliding reactants, as shown in Figure 6.5

Figure 6.5 Reaction coordinate (Potential energy diagram)

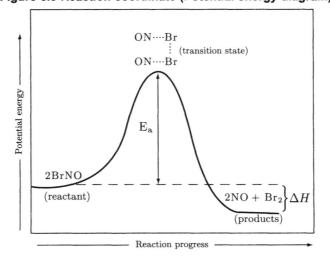

Rate law and reaction order

Rate Law   mathematical expression that links the rate of a reaction to the concentrations of the substances that influence the rate of that reaction

> Rate $= k[A]^y[B]^z$, where k is the rate constant and y and z are exponents that specify the order for reactants A and B, respectively

Rate Constant k, the proportionality constant in the rate law for a given reaction; k changes with temperature

Reaction order   the exponent on a given concentration term within the rate law; a reactant is:

- 1st order — when the concentration of a reactant is doubled and the reaction rate doubles

- 2nd order — when the concentration of the reactant is doubled and the reaction rate increases by a factor of 4

- 0 order — when the concentration of reactant is changed and there is no change in the reaction rate

Overall reaction order sum of the individual orders in the rate law

Isolation experiment   kinetics experiment which isolates the effect of one particular reactant on the reaction rate; carried out by making the starting concentration of that particular reactant significantly lower than the concentrations of the other reactants

THE MATHEMATICS OF SIMPLE REACTION KINETICS

Differential rate law data:

Compare the change in initial concentration of one reactant and its effect on initial reaction rate while holding the other reactant(s) constant. Example: for the reaction hypothetical $A + B \rightarrow AB$, consider this data set.

Trial	Initial [A]	Initial [B]	Initial rate
1	x	y	1.01
2	2x	y	2.02
3	x	2y	2.02

Compare trial 1 to trial 2 and note that, although the initial concentration of B is held constant at y, the concentration of A doubles from x to 2x. This concentration change causes the reaction rate to double, so the order of reactant A is said to be 1 or first order. Comparing trials 1 and 3 reveals that, although reactant A is held constant at x, the concentration of reactant B is doubled from y to 2y. This change in concentration also causes the initial rate of reaction to double, so reactant B is also said to be first order. The overall order of the reaction is the sum of the individual orders, in this case, second order overall.

Use differential rate laws to determine order of reactants, overall order of each reaction and rate constant from experimental data; predict reaction mechanism.

Integrated Rate Law Data:

The set of rate data for this type of experiment is shown in a different way: the concentration of a reactant is presented as a function of time, thus giving an indication of the instantaneous rate of disappearance of the reactant at a given time, t. To determine the order of a reactant, plot concentration versus time and look for a straight line relationship for some function of [C] vs. t.

Use integrated rate law data to establish the graphical relationship between concentration of reactant (some function of [C]) and reaction time, t, to determine order of each reactant, overall order of reaction and rate constant from experimental data; predict reaction mechanism.

Order	Straight line relationship
0 order:	[C] vs. time
1st order:	ln[C] vs. time
2nd order:	1/[C] vs. time

Since many reactions are first order, try graphing ln[C] versus t first; then try 1/[C] versus t to investigate 2nd order; then try [C] versus t as a test of zero order.

When the linear relationship of the natural log of the concentration of reactant C versus time, then this reaction is first order in C. Other straight line relationships are given above.

The Arrhenius equation and its applications

The Arrhenius equation includes the concept of the effective collision, which is based on particles traveling with sufficient kinetic energy to break old bonds and colliding in an orientation which then allows a new bond to form. It is usually presented in a line-slope format that compares frequency of effective collisions at two different temperatures. Calculating the slope of the line thus drawn allows you to determine the **activation energy, E_a**, of the reaction. This relationship is given by the equation

$$\ln\left(\frac{k_2}{k_1}\right) = \frac{E_a}{R}\left(\frac{1}{T_1} - \frac{1}{T_2}\right) \qquad R = 8.134 \text{ J mol}^{-1} \text{ K}^{-1}$$

where k_1 refers to the rate constant at temperature T_1 and k_2 at T_2. Note well that E_a is typically given in kJ so R must be adjusted to kJ as well. All temperatures must be noted in kelvins to agree with R. You may wish to refer to this graph in your textbook.

from the TOPIC OUTLINE (website: apcentral.collegeboard.com)

III. Reactions

D. Kinetics

    1. Concept of rate of reaction

    2. Use of experimental data and graphical analysis to determine reactant order, rate constants, and reaction rate laws

    3. Effect of temperature change on rates

    4. Energy of activation; the role of catalysts

    5. The relationship between the rate-determining step and a mechanism

from the list of CHEMICAL CALCULATIONS

11. Kinetics calculations

from the list of EQUATIONS & CONSTANTS

$\ln[A]_t - \ln[A]_0 = -kt$ associated with first order kinetics

$\dfrac{1}{[A]_t} - \dfrac{1}{[A]_0} = kt$ associated with second order kinetics

$\ln k = \dfrac{-E_a}{R}\left(\dfrac{1}{T}\right) + \ln A$ the Arrhenius equation;

                                  E_a = activation energy

from the list of RECOMMENDED EXPERIMENTS

12. Determination of the rate of a reaction and its order

Multiple Choice Questions

1. Each of the following is true about a heterogeneous catalyst EXCEPT

    (A) Its presence changes the rate of a chemical reaction.
    (B) It does not undergo a permanent chemical change.
    (C) It is in the same phase as the reacting particles.
    (D) Its presence lowers the activation energy of the overall reaction.
    (E) Its presence decreases the potential energy of the activated complex.

2. Which accounts for the increase in the rate of reaction when a catalyst is added to a reaction system?

    (A) decrease in ΔH for the reaction
    (B) increase in ΔS for the reaction
    (C) increase in potential energy of the reactants
    (D) increase in potential energy of the products
    (E) decrease in potential energy of the activated complex

3. In a system where the reaction $A_{(g)} + B_{(g)} \rightarrow C_{(g)}$ is taking place, which change increases the concentration of $A_{(g)}$?

    I.   addition of $A_{(g)}$ molecules at constant temperature and volume
    II. addition of $A_{(g)}$ molecules at constant temperature and pressure
    III. addition of $A_{(g)}$ molecules at constant volume and pressure

    (A) I only
    (B) III only
    (C) II and III only
    (D) I and III only
    (E) I, II, and III

4. All of the following apply to the reaction $A_{(g)} + B_{(g)} \rightarrow C_{(g)}$ as it is carried out at constant temperature in a sealed rigid container EXCEPT

    (A) The total pressure decreases.
    (B) The rate of reaction decreases.
    (C) The entropy of the system decreases.
    (D) The number of molecules of C decreases.
    (E) The frequency of collisions between molecules A and B decreases.

5. Consider the hypothetical reaction

$$X_{(g)} + 2Y_{(g)} \rightarrow XY_{2(g)}$$

$$\frac{\Delta[Y]}{\Delta t} = -5.0 \times 10^{-2} \text{ mol L}^{-1} \text{ sec}^{-1}$$

What is the rate of formation of $XY_{2(g)}$?

(A) -5.0×10^{-2} mol L^{-1} sec^{-1}

(B) -2.5×10^{-2} mol L^{-1} sec^{-1}

(C) 1.0×10^{-2} mol L^{-1} sec^{-1}

(D) 2.5×10^{-2} mol L^{-1} sec^{-1}

(E) 5.0×10^{-2} mol L^{-1} sec^{-1}

Question 6-11: Consider the reaction and its rate law given below:

$$2A_{(g)} + B_{(g)} \rightarrow C_{(g)}$$

$$\text{Rate} = k[A]^2[B]$$

At the beginning of one trial of this reaction $[A] = 4.0$ and $[B] = 1.0$. The observed rate was 0.048 mol C L^{-1} sec^{-1}.

6. The numerical value of k, the rate constant, for this reaction is closest to

(A) 3×10^2

(B) 4×10^0

(C) 8×10^{-1}

(D) 1×10^{-2}

(E) 3×10^{-3}

7. Which is the label for k, the rate constant?

(A) mol^2 L^{-2} sec^{-1}

(B) L mol^{-2} sec^{-1}

(C) L^2 mol^{-2} sec^{-1}

(D) L^2 sec mol^{-1}

(E) L^2 sec mol^{-2}

8. When [B] becomes 0.4 mol L^{-1}, what will be the value of [A]?

(A) 0.8 mol L^{-1}

(B) 1.6 mol L^{-1}

(C) 2.8 mol L^{-1}

(D) 3.4 mol L^{-1}

(E) 3.6 mol L^{-1}

9. Which of the following describes how the rate for this trial of the reaction, at constant temperature, changes as [B] approaches 0.4 mol L^{-1}?

    (A) The rate decreases because the concentration of the products increases.
    (B) The rate remains the same because the rate constant remains the same.
    (C) The rate remains the same because the temperature remains the same.
    (D) The rate decreases because the concentration of the reactants decreases.
    (E) The rate remains the same because the energy of activation remains the same.

10. Which applies to this reaction as it proceeds at constant temperature?

    I.   The rate of the reaction decreases.
    II. The effectiveness of collisions between reactant molecules remains the same.
    III. The frequency of collisions between reactant molecules remains the same.

    (A) I only
    (B) II only
    (C) I and II only
    (D) II and III only
    (E) I, II, and III

11. Which applies to this system when its temperature increases at constant volume?

    I.   [A] decreases at a greater rate.
    II. The value for k, the rate constant, remains the same.
    III. The rate of the reaction increases.

    (A) I only
    (B) II only
    (C) I and III only
    (D) II and III only
    (E) I, II, and III

    _____

12. All of the following increase the rate of a reaction involving a solid EXCEPT

    (A) adding more of the solid
    (B) increasing the concentration of the solid
    (C) increasing the temperature
    (D) adding a catalyst
    (E) increasing the surface area of the solid

Questions 13–17: $5Br^-_{(aq)} + BrO_3^-_{(aq)} + 6H^+_{(aq)} \rightarrow 3Br_{2(\ell)} + 3H_2O_{(\ell)}$

The reaction between bromide ions and bromate ions in acidic water solution occurs according to the equation above. The rate law for this reaction is known to be

$$Rate = k[Br^-][BrO_3^-][H^+]^2$$

13. One proposed reaction mechanism has three steps up to and including the slow step. The first step in the reaction mechanism is

$$Br^- + H^+ \rightarrow Intermediate_1$$

Which statement(s) must be true about this first step in the reaction mechanism?

    I.   Its coefficients of the reactants must correspond to exponents on those terms in the rate law.

    II.   No further step can include $Intermediate_1$ as a reactant.

    III.   A subsequent step must include BrO_3^- as a reactant.

(A) I only
(B) II only
(C) III only
(D) I and II only
(E) I and III only

14. The overall order for this reaction is

(A) 2
(B) 3
(C) 4
(D) 6
(E) 12

15. What is the effect of increasing $[H^+]$ in this reaction system at constant temperature?

(A) The value of the rate constant increases.
(B) The potential energy of the products decreases.
(C) The potential energy of the activated complex decreases.
(D) The number of collisions between H^+ and Br^- ions increases.
(E) The effectiveness of collisions between H^+ and Br^- ions increases.

16. What is the effect of adding $Br_{2(\ell)}$ to the system? (Assume negligible change in volume.)

    I.   The mass of the system increases.

    II.   The rate of reduction of BrO_3^- increases.

    III.   The rate of oxidation of Br^- decreases.

(A) I only
(B) II only
(C) I and II only
(D) II and III only
(E) I, II, and III

17. Which change will cause a decrease in the rate of the reaction?

      I.   addition of OH^- ions

     II.   removal of H^+ ions

   III.   addition of H_2O molecules

(A) II only
(B) III only
(C) I and III only
(D) II and III only
(E) I, II, and III

---

18. Consider the hypothetical reaction below taking place in a piston at constant temperature. The reaction is started by placing $A_{(g)}$ and $B_{(g)}$ in the piston. No chemical equilibrium is established.

$$A_{(g)} + 2B_{(g)} \rightarrow AB_{2(g)}$$

Which change causes an increase in the partial pressure of $B_{(g)}$?

      I.   addition of $B_{(g)}$ at constant volume

     II.   decrease in the volume of the reaction vessel

   III.   addition of $AB_{2(g)}$ at constant volume

(A) I only
(B) II only
(C) I and II only
(D) II and III only
(E) I, II, and III

19. Which applies to any reaction mechanism?

      I.   It is a list of steps that produce the overall chemical reaction.

     II.   It includes only unimolecular steps.

   III.   It cannot include reaction intermediates.

(A) I only
(B) II only
(C) III only
(D) I and II only
(E) I and III only

Questions 20-22: Consider the gas reaction

$$2X_{(g)} + Y_{(g)} \rightarrow Z_{(g)}$$

The graph below represents the changes in concentration of reactants and products between the start of reaction and Time 1.

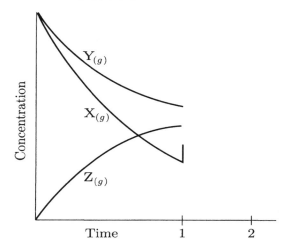

20. Which event is most likely to have occurred at Time 1?

    (A) addition of X
    (B) addition of Y
    (C) addition of Z
    (D) removal of Y
    (E) removal of Z

21. Which gives the rate of changes in concentrations that occur as the reaction proceeds from Time 1 to Time 2?

	[X]	[Y]	[Z]
(A)	decreases	decreases	decreases
(B)	decreases	increases	increases
(C)	decreases	decreases	increases
(D)	increases	increases	decreases
(E)	increases	increases	increases

22. When the lines for X, Y and Z are extended to time 2, which feature of those extensions is most closely related to the coefficients of the reactants, X and Y, in the equation for the reaction?

    (A) [Y]

    (B) [Z]

    (C) Δ[Z]

    (D) $\dfrac{\Delta[X]}{\Delta[Y]}$

    (E) [X] + [Y]

23.   Which energy distribution diagram represents an increase in temperature?

(A)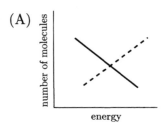

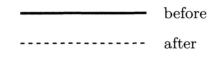 before

----------- after

(B)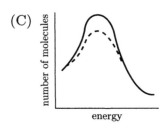

(C)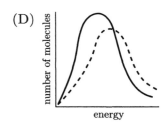

(D)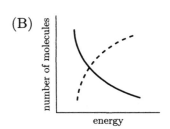

(E)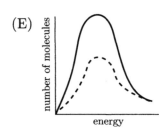

24. Which pair of potential energy/reaction coordinate diagrams represents a comparison between the same reactants where one of the pair of diagrams represents a situation that includes a catalyst?

(A)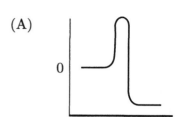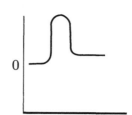

(B)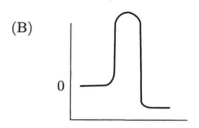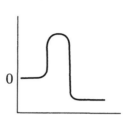

(C)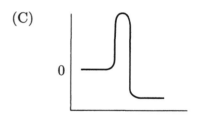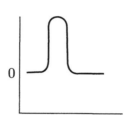

(D)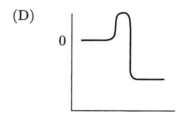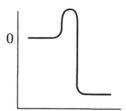

(E)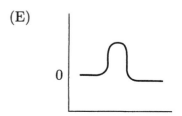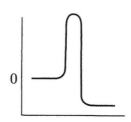

25. Which pair of potential energy/reaction coordinate diagrams represents a comparison between a reaction with high activation energy to a reaction with lower activation energy but the same heat of reaction?

(A)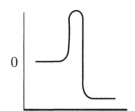

(B)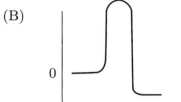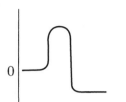

(C)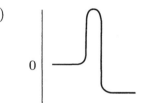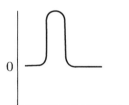

(D)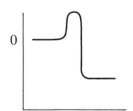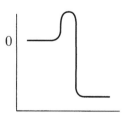

(E)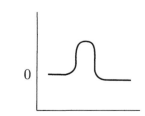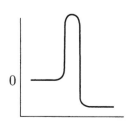

Free-Response Questions

26.
$$2A_{(g)} + B_{(g)} \rightarrow C_{(g)}$$

Trial	Initial concentration mol L^{-1} [A]	Initial concentration mol L^{-1} [B]	Initial rate of formation of C mol L^{-1} sec^{-1}
I	0.40	0.20	8.0×10^{-4}
II	0.80	0.40	1.6×10^{-3}
III	0.80	0.80	3.2×10^{-3}
IV	0.60	0.60	2.4×10^{-3}
V	0.30	?	4.0×10^{-4}

(A) Write the rate law for the reaction above in the form $Rate = k[A]^x[B]^y$ including numerical values for x and y. Explain how you determined the values for exponents x and y.

(B) Calculate the specific rate constant, k. Specify the units for k.

(C) Calculate the rate of formation of C in trial IV after [A] has decreased to 0.30 M.

(D) Calculate the initial concentration of reactant B in trial V.

(E) If the temperature were raised by 10°C for any trial, what would be the effect on the initial rate of formation of C? Explain.

27. In the reaction below, the forward reaction is known to be first order in both PCl_3 and Cl_2. The reverse reaction is first order in PCl_5.

$$PCl_{3(g)} + Cl_{2(g)} \rightleftharpoons PCl_{5(g)} + \text{energy}$$

(A) Write the rate law for the forward reaction. What factors determine the numerical values of the rate constant and of each exponential term in the rate law?

(B) Propose a mechanism for the forward reaction that is based on the molecular collision theory and describe each event (elementary process) in that mechanism.

(C) On the axes provided, draw the reaction coordinate for the forward reaction that shows relative energies of components of the system. Locate and label the relative energies of the reactants, the product and the activated complex.

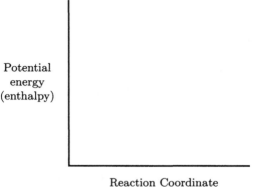

Potential
energy
(enthalpy)

Reaction Coordinate
(time, progress of the reaction)

(D) A reaction mixture is prepared in a rigid container held at constant temperature that originally contains 3.0 moles of PCl_3 and 1.0 mole of Cl_2. On the axis provided, draw a graph that shows how the quantities of Cl_2 and PCl_5 change with time until equilibrium has been established. Equilibrium is established when the quantity of PCl_5 has become 0.60 mole. The corresponding change for PCl_3 is given. Include a separate, labeled line on your graph for each of the other two substances showing how the quantity present changes over time.

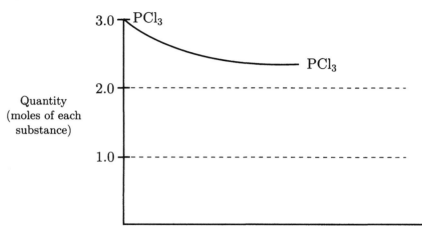

Quantity
(moles of each
substance)

Time

28.
$$H_{2(g)} + I_{2(g)} \rightarrow 2HI_{(g)}$$

For the exothermic reaction represented above, carried out at 298 K, the rate law is

$$Rate = k[H_2][I_2]$$

Predict the effect of each of the following changes on the initial rate of the reaction and explain your prediction.

(A) Removal of iodine gas at constant volume and temperature.

(B) Increase in volume of the reaction vessel at constant temperature.

(C) Addition of a catalyst.

In your explanation, include a potential energy diagram with axis labeled as potential energy versus reaction coordinate for this reaction. Show a comparison between the catalyzed and non-catalyzed reactions.

(D) Increase in temperature.

In your explanation, include a diagram, showing the number of molecules as a function of energy. Use a 'before' and 'after' presentation of information to show how the temperature change affects the distribution of molecules.

(E) Increase in overall pressure by addition of argon at constant volume.

CHAPTER 7
CHEMICAL EQUILIBRIUM

Chemical equilibrium is the single most important topic on the AP exam. It gets the most ink from the question writers. As a result, it should get the most "ink" back from students. Our best advice: you should spend any available free time improving your understanding of chemical equilibrium!

WHAT IS CHEMICAL EQUILIBRIUM?

Recognizing and explaining chemical equilibrium

In Chapter 5, we discussed reactions that go to completion; that is, reactions in which the limiting reactant is consumed and a maximum quantity of product is formed. However, in actual practice, many reaction systems reach a condition in which some quantity of each reactant remains in contact with some quantity of each product and that no further change appears to occur. The system has achieved a steady state, called equilibrium, with no apparent further change in its properties including color, mass, density and pH. Any system at equilibrium is always a closed system.

Equilibrium is recognized by constant macroscopic properties explained in terms of dynamic molecular behavior. This equilibrium condition occurs as the result of two opposing reactions that are occurring at the same rate. The rate of the forward reaction is equal to the rate of the reverse reaction. The concept of molecular collisions as the basis for chemical reaction helps explain why equilibrium exists. In the very familiar Haber process

$$N_2 + 3H_2 \rightleftharpoons 2NH_3$$

equilibrium is achieved when the rate of formation of ammonia is equal to the rate of consumption of ammonia.

The Reaction Quotient Experiments have shown that there is a reaction quotient (mass action expression) that describes quantitatively the contents of a reaction system which has reached equilibrium.

$$Q = \frac{[\text{Products}]}{[\text{Reactants}]}$$

For the generalized reaction system at equilibrium $aA + bB \rightleftharpoons cC + dD$.

$$Q = K_{eq} = \frac{[C]^c[D]^d}{[A]^a[B]^b}$$

The numerical value is known as the equilibrium constant, K_{eq}. There is a specific K_{eq} for any equilibrium system. The value of K_{eq} varies with temperature.

FACTORS AFFECTING CHEMICAL EQUILIBRIUM

Equilibrium can be disturbed when one or more of its characteristics is subjected to change.

Among the factors affecting equilibrium are

- concentration for solution systems

- volumes, concentrations or partial pressures for gas phase systems

- temperature for any system in general.

Le Chatelier's Principle By the end of the 19th Century, study of equilibrium systems was sufficiently advanced to allow the establishment of Le Chatelier's Principle, a basis for the prediction of the effects of changes in systems at equilibrium.

> **When a system at equilibrium is subjected to a stress, the system will shift so as to relieve the stress.**

AP exam questions are often presented so that students are expected to discuss equilibrium systems and changes in those systems from one or more of three perspectives:

- rates of opposing reactions (the forward and reverse reactions)

- the equilibrium constant (or reaction quotient)

- Le Chatelier's principle

SOLVING EQUILIBRIUM PROBLEMS

Quantitative equilibrium problems

Some questions call for calculations about systems that have achieved equilibrium and systems that are moving to an equilibrium position. These questions require use of K_{eq} in a wide variety of calculations. Some commonly encountered types of equilibrium systems with examples are listed below.

Gas Phase (homogeneous) Equilibrium System

$$4HCl_{(g)} + O_{2(g)} \rightleftharpoons 2H_2O_{(g)} + 2Cl_{2(g)} \qquad K_p = \frac{P_{H_2O}^2 \, P_{Cl_2}^2}{P_{HCl}^4 \, P_{O_2}}$$

Gas/Solid Phases (heterogeneous) Equilibrium System - Dissociation of a Solid

$$CaCO_{3(s)} \rightleftharpoons CaO_{(s)} + CO_{2(g)} \qquad\qquad K_p = P_{CO_2}$$

Solid/solution phases (heterogeneous) - Solubility Equilibrium

$$Ag_2CrO_{4(s)} \rightleftharpoons 2Ag^+{}_{(aq)} + CrO_4^{2-}{}_{(aq)} \qquad\qquad K_{sp} = [Ag^+]^2[CrO_4{}^{2-}]$$

Solution phase (homogeneous) equilibrium - Ionization of a Weak Acid

$$HNO_{2(aq)} + H_2O \rightleftharpoons H_3O^+_{(aq)} + NO_2^-_{(aq)} \qquad K_a = \frac{[H_3O^+][NO_2^-]}{[HNO_2]}$$

Ionization of a Weak Base

$$CH_3NH_{2(aq)} + H_2O \rightleftharpoons CH_3NH_3^+_{(aq)} + OH^-_{(aq)} \qquad K_b = \frac{[CH_3NH_3^+][OH^-]}{[CH_3NH_2]}$$

Instability   (Dissociation of a Complex Ion)

$$Ag(NH_3)_2^+_{(aq)} \rightleftharpoons Ag^+_{(aq)} + 2NH_{3(aq)} \qquad K_{inst} = \frac{[Ag^+][NH_3]^2}{[Ag(NH_3)_2^+]}$$

Dissolving/Complex Ion Formation   (Complexation)

$$Cu(OH)_{2(s)} + 4NH_{3(aq)} \rightleftharpoons Cu(NH_3)_4^{2+}_{(aq)} + 2OH^-_{(aq)}$$

$$K_{diss} = \frac{[Cu(NH_3)_4^{2+}][OH^-]^2}{[NH_3]^4}$$

Gas/Liquid heterogeneous equilibrium (Vapor Pressure)

$$\cdot\ H_2O_{(\ell)} \rightleftharpoons H_2O_{(g)} \qquad\qquad\qquad K_p = P_{H_2O}$$

Equilibrium constant expressions include only those terms whose concentrations can change such as pressures or concentrations of gases and concentrations of ions in solution.

Systems at equilibrium

For systems at equilibrium, the chemical equation is generally known, as well as enough components of the reaction quotient to permit calculation of other quantities. Solution of these problems calls for writing the equilibrium constant expression (reaction quotient or mass action expression), substituting the known quantities, then solving for the other values.

Systems moving to equilibrium

Some systems move from a previous non-equilibrium condition to a new equilibrium condition. Solution of such a problem calls for application of the principles of reaction stoichiometry to solve for concentrations at equilibrium, then further calculations using the equilibrium concentrations as determined.

One strategy commonly presented in textbooks recommends the use of a table such as that in Figure 7.1 to summarize the behavior of the system as it moves to equilibrium. Sometimes these are called "Rice", "Ice" or "Nice" tables. Especially helpful is the explicit statement of changes in quantities, Δn, or $\Delta mol\ due\ to\ rxn$, as the reaction proceeds. You should express all amounts in moles rather than moles per liter in order to avoid losing track of volume effects.

Figure 7.1 A Problem Solving Format for Equilibrium Problems
        using n_{av} Δn_{rxn} n_{av} n_{eq} $[\ ]_{eq}$: an improvement on "RICE" or "ICE"

- Use this format when a reaction occurs in a system (apply principles of stoichiometry)

  AND

- that system establishes a new equilibrium (apply principles of equilibrium).

Substance	A	B	C	D
n_{av}, mol available				
Δn, Δn_{rxn}, Δmol due to rxn				
n_{eq}, mol at equilibrium				
$[\ ]_{eq}$, conc. at equilib.				

The Students Solution Manual has many illustrations of the use of a RICE table.

Solubility equilibrium

Solubility equilibrium can be established by dissolving the solid, usually an ionic solid, into the solvent, usually water. This is generally regarded as the forward reaction. Solubility equilibrium can also be established by mixing solutions of ions that form a precipitate, in the reverse reaction. Refer to the Ag_2CrO_4 solubility equilibrium equation above. In a typical problem, the information could be provided as mass (or moles) of the specified solute dissolved per unit volume of solvent (or solution) with directions to calculate the value for K_{sp}. Alternatively, the given information may include the K_{sp}, with the molar or mass solubility as the value to be calculated.

Acid Base equilibrium

In Chapter 9, more attention will be given to the implications of chemical equilibrium for acid/base systems that include

   a.   proton transfer (donation/acceptance); K_a and K_b

   b.   self-ionization of water; K_w

   c.   ionization of weak acids and bases; K_a and K_b

   d.   ionization of polyprotic acids; K_I, K_{II} and K_{III}

   e.   hydrolysis of salts; K_h

   f.   buffer solutions

   g.   titrations/pH curves

Graphic representations

Figure 7.2 shows a plot of concentration *vs* time (progress of the reaction) as a system moves from some starting conditions and achieves equilibrium after some reactants have been consumed and some products formed. Note that an unchanging horizontal line indicates the steady state characteristic of equilibrium. An abrupt vertical shift indicates a "stress" in the form of addition or removal of some portion of one or more of the reactants or products as shown in questions 21-24 below.

Figure 7.2 Establishing equilibrium

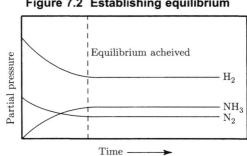

 from the TOPIC OUTLINE (website: apcentral.collegeboard.com)

III. Reactions

C. Equilibrium

    1. Concept of dynamic equilibrium, physical and chemical; Le Chatelier's Principle; equilibrium constants

    2. Quantitative treatment

        a. Equilibrium constants for gaseous reactions: K_p, K_c

        b. Equilibrium constants for reactions in solution

            (1) Constants for acids and bases; pK; pH

            (2) Solubility product constants and their application to precipitation and the dissolution of slightly soluble compounds

            (3) Common ion effect; buffers; hydrolysis

from the list of CHEMICAL CALCULATIONS

8. Equilibrium constants and their applications, including their use for simultaneous equilibria

from the list of EQUATIONS & CONSTANTS

$$Q = \frac{[C]^c [D]^d}{[A]^a [B]^b} \text{ where } aA + bB \rightarrow cC + dD$$

$$K_a = \frac{[H^+][A^-]}{[HA]}$$

$$K_b = \frac{[OH^-][HB^+]}{[B]}$$

$$K_w = [OH^+][H^+] = 10^{-14} \text{ @ } 25°C$$

$$= K_a \times K_b$$

$$pH = -\log[H^+], \ pOH = -\log[OH^-]$$

$$14 = pH + pOH$$

$$pH = pK_a + \log \frac{[A^-]}{[HA]}$$

$$pOH = pK_b + \log \frac{[HB^+]}{[B]}$$

$$pK_a = -\log K_a, \ pK_b = -\log K_b$$

$$K_p = K_c(RT)^{\Delta n}, \text{ where } \Delta n = \text{moles product gas} - \text{moles reactant gas}$$

from the list of RECOMMENDED EXPERIMENTS

10. Determination of the equilibrium constant for a chemical reaction

The experiment calls for measuring the changes in concentration of a colored ion in water solution using a spectrophotometer. Using Beer's Law, absorbance can be measured and corresponding concentration calculated. This is especially useful in determining the concentration of colored ions such as Cu^{2+} and Co^{3+} where those ions are precipitated from water solution to establish solubility equilibrium. The concentration of such ions can also be estimated using the unaided eye by comparing color intensity of solutions of known concentration to the color of solutions of unknown concentrations.

Multiple Choice Questions

1. For which reaction is K_c equal to K_p?

    (A) $H_{2(g)} + S_{(s)} \rightleftharpoons H_2S_{(g)}$

    (B) $2H_2O_{(g)} \rightleftharpoons 2H_{2(g)} + O_{2(g)}$

    (C) $3H_{2(g)} + N_{2(g)} \rightleftharpoons 2NH_{3(g)}$

    (D) $H_{2(g)} + Br_{2(\ell)} \rightleftharpoons 2HBr_{(g)}$

    (E) $2NO_{2(g)} \rightleftharpoons N_2O_{4(g)}$

2. Consider the reaction

    $$SO_{2(g)} + \tfrac{1}{2}O_{2(g)} \rightleftharpoons SO_{3(g)} \qquad K_c = 49 \text{ at } 1000 \text{ K}$$

    What is the value of K_c for the reaction below?

    $$2SO_{3(g)} \rightleftharpoons 2SO_{2(g)} + O_{2(g)}$$

    (A) $\dfrac{1}{49}$

    (B) $\dfrac{1}{7}$

    (C) $\dfrac{1}{(49)^2}$

    (D) 7

    (E) $(49)^2$

3. Consider the solubility equilibrium system represented by the equation below.

    $$PbSO_{4(s)} \rightleftharpoons Pb^{2+}{}_{(aq)} + SO_4{}^{2-}{}_{(aq)}$$

    Which occurs when $K_2SO_{4(s)}$ is added to the system at constant temperature?

    (A) $[Pb^{2+}]$ increases as the mass of $PbSO_{4(s)}$ in the system increases.

    (B) $[Pb^{2+}]$ increases as the mass of $PbSO_{4(s)}$ in the system remains the same.

    (C) $[Pb^{2+}]$ decreases as the mass of $PbSO_{4(s)}$ in the system decreases.

    (D) $[Pb^{2+}]$ decreases as the mass of $PbSO_{4(s)}$ in the system increases.

    (E) $[Pb^{2+}]$ decreases as the mass of $PbSO_{4(s)}$ in the system remains the same.

4. The K_{sp} for $PbCl_2$ is 1.6×10^{-5}. Of the following choices, which is the greatest amount of $Pb(NO_3)_{2(s)}$ that can be added to 1.0 liter of 0.010 M NaCl solution at constant temperature without causing precipitation to occur? (Assume no change in volume.)

(A) 4.0×10^{-4} mol

(B) 1.5×10^{-3} mol

(C) 8.0×10^{-3} mol

(D) 4.0×10^{-2} mol

(E) 1.5×10^{-1} mol

Questions 5-9:

$$4H_{2(g)} + CS_{2(g)} \rightleftharpoons CH_{4(g)} + 2H_2S_{(g)}$$

The system below reaches equilibrium according to the equation above. A mixture of 2.50 mol $H_{2(g)}$, 1.50 mol $CS_{2(g)}$, 1.50 mol $CH_{4(g)}$ and 2.00 mol $H_2S_{(g)}$ is placed in a 5.0 L rigid reaction vessel. When equilibrium is achieved, the concentration of $CH_{4(g)}$ has become 0.25 mol L^{-1}. The five questions below apply to this equilibrium system.

(You may use the tabular format below to help you analyze this problem. It is unlikely that this format will appear on the examination. However, its use will help you work through some quantitative principles of chemical equilibrium. Values printed in the table are those given in the problem.)

	$H_{2(g)}$	$CS_{2(g)}$	$CH_{4(g)}$	$H_2S_{(g)}$
mol available	2.50	1.50	1.50	2.00
mol change				
mol at equilibrium				
concentration mol L^{-1}			0.25	

5. Changes in concentration occur as this system approaches equilibrium. Which expression gives the best comparison of the changes in those concentrations shown in the ratio below?

$$\frac{\Delta[H_2S]}{\Delta[CS_2]}$$

(A) $\dfrac{+2}{+1}$

(B) $\dfrac{+2}{-1}$

(C) $\dfrac{-2}{+1}$

(D) $\dfrac{-1}{+1}$

(E) $\dfrac{-1}{+2}$

6. What is the change in the number of moles of $H_2S_{(g)}$ present as the system moves from its original state to the equilibrium described?

    (A) −1.25
    (B) −0.50
    (C) −0.25
    (D) +0.25
    (E) +0.50

7. What is the number of moles of $CS_{2(g)}$ at equilibrium?

    (A) 0.25
    (B) 0.35
    (C) 0.75
    (D) 1.25
    (E) 1.75

8. What is the concentration in moles per liter of $H_{2(g)}$ at equilibrium?

    (A) 0.50
    (B) 0.70
    (C) 1.0
    (D) 2.0
    (E) 3.0

9. Which correctly describes the values for ΔG, the free energy change, and Q, the mass action expression (reaction quotient), when the mixture was prepared?

    (A) $\Delta G = 0$, $Q = K_{eq}$
    (B) $\Delta G > 0$, $Q < K_{eq}$
    (C) $\Delta G > 0$, $Q > K_{eq}$
    (D) $\Delta G < 0$, $Q > K_{eq}$
    (E) $\Delta G < 0$, $Q < K_{eq}$

Questions 10–13:

$$\text{heat} + N_2O_{4(g)} \rightleftharpoons 2NO_{2(g)}$$

Consider an equilibrium system based on the reaction above. This equilibrium mixture is contained in a piston.

10.   Which occurs when the volume of the system is increased at constant temperature?

	number of molecules of NO_2	total number of molecules of all gases	K_p
(A)	increases	decreases	remains the same
(B)	increases	remains the same	remains the same
(C)	increases	increases	remains the same
(D)	remains the same	decreases	decreases
(E)	remains the same	increases	decreases

11.   Which occurs when the force on the piston is decreased at constant temperature?

	partial pressure of of N_2O_4	total pressure of of all gases	K_p
(A)	decreases	increases	remains the same
(B)	increases	decreases	remains the same
(C)	decreases	decreases	remains the same
(D)	increases	decreases	decreases
(E)	increases	increases	decreases

12.   Which occurs when the temperature of the system is increased at constant volume?

	number of molecules of N_2O_4	total number of molecules of all gases	K_p
(A)	increases	decreases	remains the same
(B)	decreases	increases	remains the same
(C)	increases	decreases	increases
(D)	decreases	decreases	increases
(E)	decreases	increases	increases

13. Which occurs when $NO_{2(g)}$ is added to the system at constant volume and temperature?

	number of molecules of N_2O_4	total number of molecules of all gases	K_p
(A)	increases	decreases	remains the same
(B)	increases	increases	remains the same
(C)	decreases	increases	remains the same
(D)	decreases	decreases	decreases
(E)	increases	increases	decreases

14. $$HgO_{(s)} + 4I^-_{(aq)} + H_2O \rightleftharpoons HgI_4^{2-}_{(aq)} + 2OH^-_{(aq)}$$

Consider the equilibrium above. Which of the following changes will increase the concentration of $HgI_4^{2-}_{(aq)}$?

    I.   adding $6M$ HNO_3

    II. increasing mass of $HgO_{(s)}$ present

   III. adding $KI_{(s)}$

(A) II only
(B) III only
(C) I and III only
(D) II and III only
(E) I, II, and III

15. The molar solubility of Ag_2CrO_4 is 1.3×10^{-4} at 25°C. Which expression gives the value for the solubility product constant, K_{sp}, for Ag_2CrO_4?

(A) $(1.3 \times 10^{-4})^2$

(B) $(1.3 \times 10^{-4})^3$

(C) $(2.6 \times 10^{-4})(1.3 \times 10^{-4})$

(D) $(2.6 \times 10^{-4})^2(1.3 \times 10^{-4})$

(E) $\dfrac{(2.6 \times 10^{-4})^2}{(1.3 \times 10^{-4})}$

Questions 16–19:

$$N_{2(g)} + 3H_{2(g)} \rightleftharpoons 2NH_{3(g)} + 92 \text{ kJ}$$

The four questions below apply to an equilibrium system based on the reversible reaction given above.

16. When the temperature of such an equilibrium system is increased at constant volume, which property is least affected?

    (A) density
    (B) pressure
    (C) concentration of $NH_{3(g)}$
    (D) concentration of $N_{2(g)}$
    (E) average kinetic energy

17. How are the rates of the opposing reactions and the value of the equilibrium constant, K_p, affected when heat is added to this system?

    (A) The reaction rates remain unchanged and K_p decreases.
    (B) The forward reaction will be favored and K_p decreases.
    (C) The forward reaction will be favored and K_p increases.
    (D) The reverse reaction will be favored and K_p decreases.
    (E) The reverse reaction will be favored and K_p increases.

18. Which observation confirms the fact that equilibrium has been reached in such a system confined in a closed, rigid container?

    (A) The density remains constant.
    (B) The odor of ammonia can first be detected.
    (C) The pressure is decreasing at a constant rate.
    (D) The partial pressure of hydrogen remains constant.
    (E) The mass of the system has decreased to a constant value.

19. Which occurs when such an equilibrium system is subjected to a stress by the addition of $H_{2(g)}$ and the system proceeds to a new equilibrium?

    I.   Heat will be released.
    II. Some of the added $H_{2(g)}$ will be consumed.
    III. Some of the $N_{2(g)}$ present originally will be consumed.

    (A) II only
    (B) III only
    (C) I and III only
    (D) II and III only
    (E) I, II, and III

20. Consider the equilibrium system

$$A_{(g)} + B_{(g)} \rightleftharpoons C_{(g)} + D_{(g)} + heat$$

What is the effect on the rates of the opposing reactions when the temperature of this system is increased?

(A) The $rate_{forward}$ increases; the $rate_{reverse}$ decreases.
(B) The $rate_{forward}$ decreases; the $rate_{reverse}$ decreases.
(C) The $rate_{forward}$ increases; the $rate_{reverse}$ remains the same.
(D) The $rate_{forward}$ remains the same; the $rate_{reverse}$ decreases.
(E) Both $rate_{forward}$ and $rate_{reverse}$ increase.

Questions 21-24: Consider the reaction

$$CO_{(g)} + Cl_{2(g)} \rightleftharpoons COCl_{2(g)} + heat$$

A system is established in which the original concentration of CO is 10.0 M and that of Cl_2 is 8.0 M. The system is contained in a sealed rigid container. The progress of the reaction is given in the graph below. The system is maintained at constant temperature.

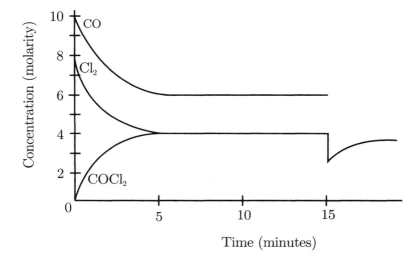

21. Which is the best comparison of Q, the reaction quotient, to K_c, the equilibrium constant, at time 2.5 minutes?

(A) $Q > 0$ and $Q > K_c$
(B) $Q > 0$ and $Q = K_c$
(C) $Q > 0$ and $Q < K_c$
(D) $Q < 0$ and $Q > K_c$
(E) $Q < 0$ and $Q < K_c$

22. Which is the best description of the rate of the forward reaction, Rate$_{forward}$, between time 0 and time 4 minutes?

(A) Rate$_{forward}$ > Rate$_{reverse}$ and decreasing

(B) Rate$_{forward}$ > Rate$_{reverse}$ and increasing

(C) Rate$_{forward}$ = Rate$_{reverse}$ and constant

(D) Rate$_{forward}$ < Rate$_{reverse}$ and decreasing

(E) Rate$_{forward}$ < Rate$_{reverse}$ and increasing

23. Which is the best explanation of the cause of the changes in concentration that occur after time 15 minutes?

(A) increase in temperature

(B) decrease in temperature

(C) addition of a suitable catalyst

(D) addition of some $COCl_2$

(E) removal of some $COCl_2$

24. Which is the best comparison of the

- rate of the forward reaction at 10 minutes, Rate$_{forward,10\ min}$

   and the

- concentration of CO at 10 minutes, $[CO]_{10\ min}$, the first equilibrium

   to the

- corresponding characteristics at 20 minutes, when the second equilibrium is established

(A) Rate$_{forward,10min}$ = Rate$_{forward,20min}$; with $[CO]_{10\ min}$ = $[CO]_{20\ min}$

(B) Rate$_{forward,10min}$ < Rate$_{forward,20min}$; with $[CO]_{10\ min}$ = $[CO]_{20\ min}$

(C) Rate$_{forward,10min}$ > Rate$_{forward,20min}$; with $[CO]_{10\ min}$ = $[CO]_{20\ min}$

(D) Rate$_{forward,10min}$ < Rate$_{forward,20min}$; with $[CO]_{10\ min}$ > $[CO]_{20\ min}$

(E) Rate$_{forward,10min}$ > Rate$_{forward,20min}$; with $[CO]_{10\ min}$ > $[CO]_{20\ min}$

25. Calcium fluoride, CaF_2, is a nearly insoluble salt with $K_{sp} = 3.9 \times 10^{-11}$ at 298 K. Addition of which substance is most likely to increase the solubility of CaF_2 at constant temperature?

(A) $NaF_{(s)}$

(B) $CaCl_{2(s)}$

(C) $HCl_{(aq)}$

(D) H_2O

(E) $C_2H_5OH_{(\ell)}$

Free-Response Questions

26. $$PCl_{5(g)} \rightleftharpoons PCl_{3(g)} + Cl_{2(g)}   \quad K_p = 630 \text{ at } 546 \text{ K}$$

A system is prepared by placing equimolar amounts of the three gases shown in the equation above in a suitable rigid container held at constant volume. Equilibrium is established at 546 K.

(A) When equilibrium is established, how does $[PCl_3]$ compare to $[Cl_2]$? Explain.

(B) When equilibrium is established, how does $[PCl_5]$ compare to $[PCl_3]$? Explain.

(C) When equilibrium is established, how does the rate of the forward reaction compare to the rate of the reverse reaction? Explain.

(D) If the volume of the container is increased at constant temperature of 546 K, the equilibrium is disrupted.

    (1) What effect is observed on the number of moles of $PCl_{5(g)}$ in the system? Explain.

    (2) Does the value of K_p increase, decrease or remain the same? Explain.

27. Calcium sulfate, $CaSO_4$ (molar mass: 136 g), is a nearly insoluble salt with a solubility product constant, K_{sp}, of 2.4×10^{-5}.

(A) Write the chemical equation for the solubility equilibrium of calcium sulfate.

(B) Calculate the solubility of calcium sulfate in grams solute per 100 grams of solvent. Assume that the volume occupied by the solute in the solution is negligibly small.

(C) Does precipitation occur when 100. mL of 0.050 M $CaCl_{2(aq)}$ is added to 100. mL of 0.10 M $K_2SO_{4(aq)}$? Show calculations to support your response.

(D) In another experiment, 100. mL of 0.020 M $CaCl_{2(aq)}$ is added to 200. mL of 0.20 M $K_2SO_{4(aq)}$. Calculate the equilibrium concentrations of $Ca^{2+}{}_{(aq)}$ and $SO_4{}^{2-}{}_{(aq)}$.

28. A system is prepared by placing 2.00 mol $H_{2(g)}$ and 3.00 mol $I_{2(g)}$ in a 3.00 L container at 700. K. The system reaches equilibrium according to the equation

$$H_{2(g)} + I_{2(g)} \rightleftharpoons 2HI_{(g)}$$

The number of moles of $HI_{(g)}$ at equilibrium is 1.90.

(A) Write the equilibrium expression, K_c, for this system.

(B) Calculate the value for K_c.

(C) Calculate the partial pressure in atmosphere of $I_{2(g)}$ at equilibrium.

(D) If the system were changed by adding 1.00 mol $H_{2(g)}$, what would be the partial pressure of $HI_{(g)}$ at the new equilibrium?

CHAPTER 8
THERMODYNAMICS

Another major topic for the Advanced Placement Chemistry examination is that of thermodynamics. You should be familiar with problems that trace heat flow (enthalpy). You should also understand the role of entropy and entropy changes, as well as the trend in the universe towards increasing disorder. Changes in enthalpy and entropy are linked in the Gibbs Free Energy equation which helps identify spontaneous processes.

THERMODYNAMICS – THREE LAWS

Thermodynamics is the study of energy and how it can be expressed as heat or work. The **First Law of Thermodynamics** states that the amount energy in the universe is a finite quantity and that it cannot be created nor destroyed. It can be expressed as either heat or as work. In chemistry, we frequently assume that no work is being performed (since ions or molecules in solution do not exert a force over a distance). You can track energy changes by observing only heat flow (**changes in enthalpy**) in relation to chemical reactions. In this branch of thermodynamics, called **thermochemistry**, you usually make measurements of heat flow using a **calorimeter**. The **coffee cup calorimeter** is made by stacking together several cups made of expanded polystyrene (hence the tradename "Styrofoam"). Any reaction that occurs in a solution contained within this well-insulated container is assumed to retain all heat that is produced. No heat is exchanged with the immediate surroundings. It is also assumed that any reaction occurs at constant pressure. While these assumptions may not always be true, they are frequently close enough to allow making reasonable temperature measurements and corresponding heat flow calculations. If a constant volume, variable pressure situation is to be studied, a **bomb calorimeter** can be used instead. For this device, the heat absorbed by the calorimeter and its contents is carefully measured and factored into any calculations. **Hess's Law** allows you to predict the enthalpy changes of a reaction by combining any plausible (or measurable) set of equations for chemical reactions that lead to the net reaction under consideration. The enthalpy changes, ΔH, for these component reactions are obtained from a reference source or measured in the laboratory, then combined algebraically. You can use this method because enthalpy is a **state function**, a characteristic whose value is independent of the pathway followed to reach it.

The **Second Law of Thermodynamics** specifies that in any spontaneous process, the entropy S (or disorder) of the universe increases. The effect of **entropy** changes on chemical reactions is much more subtle than that of enthalpy changes, yet it can have a measurable effect in certain circumstances as noted in the equation that defines **Gibbs Free Energy** change, $\Delta G = \Delta H - T\Delta S$. Calculation of ΔG permits prediction of direction of progress for a chemical reaction; that is, whether or not a reaction is spontaneous. A negative value for ΔG indicates a spontaneous reaction, or that the forward reaction is favored. A positive value predicts a non-spontaneous forward reaction, or that the reverse reaction is spontaneous. A value of zero indicates that the forward reaction is as favorable as the reverse reaction, or that an equilibrium position has been attained.

The **Third Law of Thermodynamics** includes a definition that establishes a baseline for entropy values. The entropy of a perfectly aligned solid crystal at zero kelvins (K) is assumed to be zero since there can be no situation more ordered.

The Three Laws of Thermodynamics

First Law of Thermodynamics
energy cannot be created nor destroyed, just transferred - energy in the universe is conserved: E(energy) = Q(heat) + W(work)

Second Law of Thermodynamics
the total entropy of the universe increases in a spontaneous process; entropy is always increasing in the universe

Third Law of Thermodynamics
a pure, perfect crystal at 0 kelvins has zero entropy

spontaneous
the direction a process will take, if left alone and given sufficient time; indicated by sign of ΔG value; spontaneity is a state function

entropy (S)
the quantitative measure of disorder or randomness, influenced by temperature and position of reacting particles;

$$\Delta S = \sum S_{products} - \sum S_{reactants}$$

$$\Delta G = \Delta H - T\Delta S$$

At equilibrium $\Delta G = 0$, therefore $\Delta S = \Delta H/T$

standard thermodynamic conditions
298 K, 1 atm, 1 M

free energy change (ΔG)
the part of a system's energy that is ordered and available to become spontaneously disordered; the balance struck between the tendency toward minimum energy and the tendency toward maximum disorder

Functions and Systems

state function
function or property whose value depends only on the present state of the system, not on the path used to arrive at the condition.

path function
function that depends upon how the change occurs; not frequently addressed in AP chemistry

system
the material(s) and their changes being studied and measured

open system
both matter and energy can be exchanged between system and surroundings

closed system
energy can be exchanged between components of the system, only; no exchange of matter

isolated or insulated system
neither energy nor matter can be exchanged

Temperature, Heat and Work

temperature	a measure of the average kinetic energy associated with molecular motion
heat (Q)	transfer of thermal energy between a system and its surroundings, flows from hot to cold
molar heat capacity (C_p)	amount of heat energy required to raise the temperature of one mole of substance by one degree Celsius at constant pressure
work	energy used to move an object against an opposing force
pressure	force/unit area; measured in kPa, atm, torr, or mmHg
Enthalpy change (ΔH)	heat (energy change) for a process under constant pressure conditions; predict ΔH for a reaction by using standard heats of formation or bond energies.

$$\Delta H° = \sum \Delta H°_{f,\,products} - \sum \Delta H°_{f,\,reactants}$$
$$\Delta H_{rx} = \sum bonds\ broken - \sum bonds\ formed$$

Hess's Law of Heat of Summation	the enthalpy change for any process is equal to the sum of the enthalpy changes for any set of steps that leads from the initial to final condition of the process;

$$\Delta H_{total} = \Delta H_1 + \Delta H_2 + \Delta H_3$$

Energy unit	1 joule (J) = 1 kg m^2/sec^2 (as in $\frac{1}{2}mv^2$)

4.18 J = 1 calorie

= the amount of heat needed to raise the temperature of one gram of water by one degree Celsius

Heat (energy flow)	$Q = m\ c\ \Delta T$ describes the energy flow within a substance as heat, where

m = mass of substance;

c = specific heat capacity of the substance;

ΔT = change in temperature

calorimeter	an isolated system calibrated prior to use to determine its heat capacity;

Q lost by rx = Q gained by reaction system + Q gained by calorimeter

$$= mc\Delta T + C_{cal}\Delta T$$

Changes in Work and Enthalpy

ΔH – change in enthalpy

Endothermic – describes a system which absorbs heat energy; ΔH (+); $\Delta H > 0$

Exothermic – describes a system which releases heat energy; ΔH (−); $\Delta H < 0$

Work including Δ(PV) work as energy

(+) Work performed on a chemical reaction system from the universe; "endoworkic", a concept parallel to endothermic; eg W or $\Delta(PV)$(+); W or $\Delta(PV) > 0$

(−) Work performed by a chemical reaction system against the universe; "exoworkic" a concept parallel to exothermic; e.g. burning a fuel; W or $\Delta(PV)$(−); W or $\Delta(PV) < 0$

$$W = P_{ext}\Delta V$$

Indications of Increasing Entropy

1. For progression through the phases of matter, entropy increases; $(s) < (\ell) < (g)$. Gases are clearly more disordered than liquids or solids.

2. In the dissolving process, solutions become more entropic as solid or liquid solute is dispersed in a liquid solvent. When a gas solute dissolves in a liquid solvent, entropy decreases. The system becomes more ordered.

3. Entropy increases with increasing mass and increasing numbers of particles since there are more positions possible for atoms or electrons.

4. Entropy increases with increasing delocalization of electrons. Ionic bonds are formed due to transfer of electrons which are then permanently localized to one nucleus. In covalent bonds, electrons are shared between two nuclei. In metallic bonds, electrons are shared among nuclei. Entropy increases from ionic to covalent to metallic bonds in terms of delocalization of electrons.

5. Increasing entropy ($\Delta S > 0$) is associated with weaker bonds and increasing softness.

6. Entropy increases with chemical complexity. For example, consider the trend in entropy values from NaCl (72.33 J mol^{-1} K^{-1}) to MgCl$_2$ (89.0) to AlCl$_3$ (109.3).

Gibbs Free Energy

$\Delta G < 0$ (negative) means spontaneous (forward) reaction

$\Delta G > 0$ (positive) means non-spontaneous (forward) reaction

$\Delta G = \Delta H - T\Delta S$
    for indicated combinations of values of ΔH and ΔS respectively

$\Delta G° = \Delta H° - T\Delta S°$
    where ° refers to the standard conditions of 1 atm and 298 K.
    Note: $\Delta G°$ is read as *delta gee zero, delta gee naught* or even *delta gee halo*)

How changes in enthalpy and entropy affect Gibbs Free Energy change

Some nonspontaneous endothermic reactions become spontaneous at a sufficiently elevated temperature, for example, boiling water at 100 °C and 1 atm.

Similarly, some exothermic reactions become nonspontaneous at a sufficiently elevated temperature, for example, freezing water at 0°C and 1 atm. See Figure 8.1 for a quick summary of these thermodynamic parameters.

Figure 8.1 Thermodynamic parameters

ΔH	ΔS	ΔG for the process under consideration
$(-)$	$(+)$	Spontaneous at any temperature
$(+)$	$(+)$	Endothermic non-spontaneous reactions become spontaneous at sufficiently high temperature
$(-)$	$(-)$	Exothermic spontaneous reactions become non-spontaneous at sufficiently high temperature
$(+)$	$(-)$	Non-spontaneous at any temperature

Linking Thermodynamics, Equilibrium and Electrochemistry

A spontaneous reaction is indicated by a negative value for $\Delta G°$, a positive voltage, $E°$, and a positive exponent for a K_{eq} value.

$$\text{The quantitative link: } \Delta G° = -R\,T\,\ln K_{eq} = -n\Im E°$$

See Fig. 8.2 for a quick summary of spontaneity and its relation to ΔH and ΔS.

Figure 8.2 Summary of Thermodynamic terms

Term	Symbol	What it measures	Interpretation
Enthalpy change	ΔH	heat flow	$-$ = exothermic $+$ = endothermic
Entropy change	ΔS	disorder	$-$ = increase in disorder $+$ = decrease in disorder
Gibbs Free Energy change	ΔG	spontaneity	$-$ = process will occur $+$ = process will not occur

from the TOPIC OUTLINE (website: apcentral.collegeboard.com)

III. Reactions

E. Thermodynamics

1. State functions

2. First law: change in enthalpy; heat of formation; heat of reaction; Hess's law; heats of vaporization and fusion; calorimetry

3. Second law: entropy; free energy of formation; free energy of reaction; dependence of change in free energy on enthalpy and entropy changes

4. Relationship of change in free energy to equilibrium constants and electrode potentials

from the list of CHEMICAL CALCULATIONS

10. Thermodynamic and thermochemical calculations

from the list of EQUATIONS & CONSTANTS

$$\Delta S^\circ = \sum S^\circ_{products} - \sum S^\circ_{reactants}$$

$$\Delta H^\circ = \sum \Delta H^\circ_{f\ products} - \sum \Delta H^\circ_{f\ reactants}$$

$$\Delta G^\circ = \sum \Delta G^\circ_{f\ products} - \sum \Delta G^\circ_{f\ reactants}$$

$$\Delta G^\circ = \Delta H^\circ - T\Delta S^\circ = -RT\ \ln K = -2.303RT\ \log K = -n\Im E^\circ$$

$$q = mc\Delta T$$

$$C_p = \Delta H/\Delta T$$

c = specific heat capacity, T = temperature, K = equilibrium constant, V = volume

from the list of RECOMMENDED EXPERIMENTS

13. Determination of enthalpy change associated with a reaction

Multiple Choice Questions

Questions 1-5: The set of lettered choices is a list of symbols used in thermodynamics and thermochemistry. Select the one lettered choice that best fits each numbered statement. A choice may be used once, more than once or not at all.

(A) K_c

(B) ΔG

(C) ΔH

(D) ΔS

(E) E_a

1. Its value is negative for any exothermic reaction.

2. Its value is negative for any spontaneous reaction.

3. Its value is calculated using the absolute (Kelvin) temperature and two of the other choices in the set.

4. Its value represents the change in randomness as a reaction proceeds.

5. Its value is determined by using molar concentrations at equilibrium in the mass action expression.

Questions 6-8: $N_{2(g)} + 3H_{2(g)} \rightleftharpoons 2NH_{3(g)}$ $\Delta H° = -92$ kJ

The Haber process is a commercial method for the manufacture of ammonia. It is based upon the equilibrium shown in the equation above. The absolute standard entropies, $S°$, of the components of this system are given below.

$$S°_{N_2} = 192 \text{ J K}^{-1} \text{ mol}^{-1}$$
$$S°_{H_2} = 131 \text{ J K}^{-1} \text{ mol}^{-1}$$
$$S°_{NH_3} = 193 \text{ J K}^{-1} \text{ mol}^{-1}$$

6. The value of $\Delta S°$ in J K^{-1} for the reaction above is closest to

(A) +200

(B) +100

(C)   0

(D) −100

(E) −200

7. Which interval includes ΔG for this reaction at 200 K, assuming ΔS and ΔH are constant over a wide range of temperatures?

   (A) less than -100 kJ

   (B) between -100 and -40 kJ

   (C) between -40 and $+40$ kJ

   (D) between -40 and $+100$ kJ

   (E) greater than $+100$ kJ

8. Which describes the change in ΔG and the change in spontaneity of the reaction as the temperature increases to values above 460 K?

   (A) ΔG increases to zero and the reaction achieves equilibrium.

   (B) ΔG decreases to values less than zero and the reaction becomes spontaneous.

   (C) ΔG decreases to values less than zero and the reaction becomes non-spontaneous.

   (D) ΔG increases to values greater than zero and the reaction becomes spontaneous.

   (E) ΔG increases to values greater than zero and the reaction becomes non-spontaneous.

---

9. The molar heat of fusion, ΔH_{fus}, for water is 6.01 kJ mol^{-1}. Which expression gives the molar entropy of fusion, ΔS_{fus}, in kJ K^{-1} mol^{-1} for ice at its normal melting point?

   (A) $\dfrac{6.01}{273}$

   (B) $\dfrac{6.01}{298}$

   (C) 6.01×4.18

   (D) 6.01×273

   (E) 6.01×298

10. The molar heat of fusion, ΔH_{fus}, for water is 6.01 kJ mol^{-1}. The specific heat capacity for water, C_p, is 75 J mol^{-1} °C^{-1}. Which expression gives the quantity of energy needed to change 1.0 mol ice at 0°C to liquid water at 25°C?

   (A) $\dfrac{6010}{75 \times 25}$

   (B) $6.01 + 75$

   (C) $6010 + (75 \times 25)$

   (D) $\dfrac{6010}{298} + (75 \times 25)$

   (E) $\dfrac{6010 + (75 \times 25)}{298}$

11. The heat of neutralization for a strong acid in dilute water solution is about 60 kJ mol^{-1}. What quantity of heat in kJ is produced when 100. mL of 1.0 M H_2SO_4 is mixed with 100. mL of 1.0 M KOH?

   (A)   0.10
   (B)   0.30
   (C)   0.40
   (D)   6.0
   (E)   18

12. Which reaction illustrates the least increase in entropy?

   (A) $N_2O_{4(g)} \rightarrow 2NO_{2(g)}$
   (B) $C_6H_{6(\ell)} \rightarrow C_6H_{6(g)}$
   (C) $2KClO_{3(s)} \rightarrow 3O_{2(g)} + 2KCl_{(s)}$
   (D) $H_{2(g)} + Cl_{2(g)} \rightarrow 2HCl_{(g)}$
   (E) $C_2H_5OH_{(\ell)} + 3O_{2(g)} \rightarrow 2CO_{2(g)} + 3H_2O_{(g)}$

13. For which reaction is ΔH most nearly equal to ΔE?

   (A) $H_{2(g)} + \frac{1}{2}O_{2(g)} \rightarrow H_2O_{(g)}$
   (B) $H_{2(g)} + Cl_{2(g)} \rightarrow 2HCl_{(g)}$
   (C) $C_2H_5OH_{(\ell)} \rightarrow C_2H_5OH_{(g)}$
   (D) $N_2O_{4(g)} \rightarrow 2NO_{2(g)}$
   (E) $BaO_{2(s)} \rightarrow BaO_{(s)} + \frac{1}{2}O_{2(g)}$

14. Which equation represents the standard formation reaction for $BaSO_{4(s)}$ at 298 K?

   (A) $BaS_{(s)} + 2O_{2(g)} \rightarrow BaSO_{4(s)}$
   (B) $Ba_{(s)} + \frac{1}{8}S_{8(s)} + 2O_{2(g)} \rightarrow BaSO_{4(s)}$
   (C) $Ba_{(s)} + S_{(s)} + 4O_{(g)} \rightarrow BaSO_{4(s)}$
   (D) $Ba^{2+}{}_{(aq)} + S^{6+}{}_{(aq)} + 4O^{2-}{}_{(aq)} \rightarrow BaSO_{4(s)}$
   (E) $Ba^{2+}{}_{(aq)} + SO_4{}^{2-}{}_{(aq)} \rightarrow BaSO_{4(s)}$

15. Which describes the thermodynamic parameters for the system below at 298 K and one atm?

$$H_2O_{(g)} \rightleftharpoons H_2O_{(\ell)}$$

	ΔG	ΔH	ΔS
(A)	0	> 0	> 0
(B)	0	> 0	< 0
(C)	< 0	> 0	< 0
(D)	< 0	< 0	< 0
(E)	< 0	< 0	> 0

16.
$$CH_{4(g)} + 2O_{2(g)} \rightarrow CO_{2(g)} + 2H_2O_{(\ell)} \qquad \Delta H° = -889 \text{ kJ}$$

$$\Delta H_f° \text{ for } H_2O_{(\ell)} = -286 \text{ kJ mol}^{-1}$$
$$\Delta H_f° \text{ for } CO_{2(g)} = -393 \text{ kJ mol}^{-1}$$

Which expression gives the standard heat of formation, $\Delta H_f°$, for methane, CH_4, in kJ mol^{-1} based on the data above?

(A) $-889 + 2(286) + 393 + 2(32)$
(B) $889 - 2(286) - 393 - 2(32)$
(C) $-889 + 2(286) + 393$
(D) $889 + 2(286) + 393$
(E) $889 - 2(286) - 393$

17. A cube of ice is added to some hot water in an insulated container which is then sealed. There is no heat exchange with the surroundings. Which describes the system once it has shifted to a new equilibrium?

    I.   The average kinetic energy of the liquid phase has decreased.

    II.   The total energy of the system has decreased.

    III.   The entropy of the system has increased.

(A) I only
(B) III only
(C) I and II only
(D) I and III only
(E) I, II, and III

18. One version of the First Law of Thermodynamics is expressed as

$$\Delta E = q + w$$

Which gives the sign convention for this relationship that is usually used in chemistry?

	heat, q added to the system	heat, q added to the surroundings	work, w done on the system	work, w done on the surroundings
(A)	$-$	$+$	$-$	$-$
(B)	$+$	$+$	$+$	$+$
(C)	$+$	$+$	$+$	$-$
(D)	$+$	$-$	$+$	$-$
(E)	$+$	$-$	$-$	$-$

19.
$$C_2H_5OH_{(\ell)} + 3O_{2(g)} \rightarrow 2CO_{2(g)} + 3H_2O_{(g)}$$

At 298 K, the change in enthalpy, ΔH, for the reaction above is given by

(A) $\Delta E + RT$

(B) $\Delta E - 2RT$

(C) $\Delta E + 2RT$

(D) $5RT - \Delta E$

(E) $2RT - \Delta E$

20. Which applies to any reaction that proceeds spontaneously to form products from initial standard state conditions?

I. $\Delta G < 0$

II. $K_{eq} > 1$

III. $\Delta H < 0$

(A) I only

(B) II only

(C) I and II only

(D) II and III only

(E) I, II, and III

21.
$$C_{(s)} + O_{2(g)} \rightarrow CO_{2(g)} \qquad \Delta H^\circ_{comb} = -394 \text{ kJ}$$

$$H_{2(g)} + \tfrac{1}{2}O_{2(g)} \rightarrow H_2O_{(\ell)} \qquad \Delta H^\circ_{comb} = -286 \text{ kJ}$$

$$C_3H_{8(g)} + 5O_{2(g)} \rightarrow 3CO_{2(g)} + 4H_2O_{(\ell)} \qquad \Delta H^\circ_{comb} = -2{,}222 \text{ kJ}$$

$$3C_{(s)} + 4H_{2(g)} \rightarrow C_3H_{8(g)} \qquad \Delta H^\circ_f =$$

Using values above, which expression gives the heat of formation, $\Delta H_f{}^\circ$, for propane, C_3H_8?

(A) $-2{,}222 + [394 - 286]$

(B) $2{,}222 - 394 - 286$

(C) $-2{,}222 + [-3(394) - 4(286)]$

(D) $-2{,}222 + [3(394) + 4(286)]$

(E) $2{,}222 + [3(-394) + 4(-286)]$

22.
$$2C_{(s)} + 3H_{2(g)} \rightarrow C_2H_{6(g)}$$

The reaction above is not spontaneous at any temperature. Which applies to the system at 298 K?

	ΔG	ΔH	ΔS
(A)	+	+	+
(B)	+	+	−
(C)	−	+	+
(D)	−	−	−
(E)	−	−	+

23. Which applies to any reaction that has negative values for both ΔH and ΔS?

  I. The reaction is spontaneous at all temperatures.

  II. The reaction is exothermic.

  III. ΔG increases (becomes more positive) as temperature increases.

 (A) I only
 (B) III only
 (C) I and III only
 (D) II and III only
 (E) I, II, and III

24. For a given reaction, the values for standard free energy change, $\Delta G°$, and the equilibrium constant, K_{eq}, are both measures of the extent to which a reaction proceeds. Which is a reasonable value for $\Delta G°$ in kJ mol^{-1} when the corresponding value for $K_{eq} = 6.9 \times 10^5$ at 298 K?

 (A) -100
 (B) -30
 (C) 0
 (D) $+30$
 (E) $+100$

25. One version of the Second Law of Thermodynamics for any spontaneous process includes the statement

$$\Delta S_{universe} = (\Delta S_{system} + \Delta S_{surroundings}) > 0$$

Which is the best interpretation of this statement?

 (A) For any spontaneous process, the entropy of the universe increases.

 (B) For any spontaneous process, the free energy change is negative.

 (C) For any spontaneous process, the entropy change for the system is positive.

 (D) For any spontaneous process, the entropy change for the system is less than the entropy change for the surroundings.

 (E) For any spontaneous process, the entropy change for the system is greater than the entropy change for the surroundings.

Free-Response Questions

26. The Gibbs Free Energy change, ΔG, describes the spontaneity of a chemical change in terms of temperature and changes in enthalpy and entropy terms, as affected by temperature.

    (A) What is the value of ΔG for any reaction system at equilibrium?

    (B) What is the sign of ΔG for any reaction that is spontaneous?

    (C) What are the signs of ΔH and ΔS for each of the reaction circumstances described under the following situations? Explain.

    (1) spontaneous regardless of temperature
    (2) spontaneous at low temperatures but non-spontaneous at high temperatures
    (3) spontaneous at high temperatures but non-spontaneous at low temperatures
    (4) non-spontaneous regardless of temperature

27. The compound $BrCl_{(g)}$ is produced by mixing bromine gas and chlorine gas at 298 K. The mixture reaches equilibrium in a very short time.

$$Br_{2(g)} + Cl_{2(g)} \rightleftharpoons 2BrCl_{(g)}$$

	ΔH_f°, kJ mol^{-1}	ΔG_f°, kJ mol^{-1}	S°, J mol^{-1} K^{-1}
$Br_{2(g)}$	30.7	3.14	152.2
$BrCl_{(g)}$	14.7	−0.88	239.7
$Cl_{2(g)}$	0	0	222.8

    (A) What is the standard enthalpy of reaction, ΔH°, in kJ at 298 K, for the reaction between $Br_{2(g)}$ and $Cl_{2(g)}$ as shown above?

    (B) What is the numerical value of the standard entropy change, ΔS°, at 298 K for this reaction? Include units.

    (C) What is the numerical value of the standard entropy change of formation, ΔS_f°, for $BrCl_{(g)}$ at 298 K? Include units.

    (D) What is the numerical value of the equilibrium constant, K_p, for this reaction at 298 K?

28.

	ΔH_f°, (kJ mol^{-1})	ΔG_f°, kJ mol^{-1}	S°, J mol^{-1} K^{-1}
$CO_{2(g)}$	−393.5	−394.4	213.6
$C_{(s)}$	0	0	5.69
$H_{2(g)}$	0	0	130.58
$H_2O_{(\ell)}$	−285.83	−236.81	69.96
$O_{2(g)}$	0	0	205.0
$C_6H_{12}O_{6(s)}$	−1,273.02	?	212.1

All values taken at 298 K.

(A) Write a balanced equation for the formation of glucose, $C_6H_{12}O_{6(s)}$, from its elements. Include the symbol for the phase of each reactant and product in its standard state.

(B) Calculate the standard entropy change of formation, ΔS_f°, for glucose, $C_6H_{12}O_{6(s)}$ at 298 K. Specify units.

(C) Calculate the standard free energy change of formation, ΔG_f°, for glucose, $C_6H_{12}O_{6(s)}$ at 298 K. Specify units.

(D) At 298 K, glucose burns in oxygen to form only $CO_{2(g)}$ and $H_2O_{(\ell)}$. Calculate the expected value of the standard enthalpy of combustion, ΔH_{comb}°, for glucose, $C_6H_{12}O_{6(s)}$. Specify units.

CHAPTER 9
ACID-BASE SYSTEMS

Acids and bases are key players in the Advanced Placement Chemistry game. You should be able to use acid-base concepts to write net ionic equations, to form complex ions, and to solve stoichiometry, equilibrium, and thermodynamics problems.

ACIDS AND BASES – SOME DEFINITIONS

The designation of certain reactions as acid-base reactions is intended to help you organize information about reactions that illustrate similar properties. Three categories of acid-base reactions are included in the AP course:

- Arrhenius
- Bronsted/Lowry
- Lewis

For each category, a definition of acid and base is established and a characteristic reaction is described. See Figure 9.1.

Figure 9.1 Comparing acid-base systems

		Arrhenius	Bronsted/Lowry	Lewis
D E F I N I T I O N S	**An acid is a substance...**	whose only positive ion in water solution is H^+ ion	which acts as a proton (H^+) donor	which accepts a share in a pair of electrons
	examples	HCl, HNO_3	H_3O^+, HSO_4^-	BF_3, Fe^{3+}
	A base is a substance...	whose only negative ion in water solution is OH^- ion	which acts as a proton (H^+) acceptor	which donates a share in a pair of electrons
	examples	$NaOH$, KOH	NH_3, SO_4^{2-}	CN^-, $C_2O_4^{2-}$
	Characteristic reaction	acid + base → salt + water	$acid_1 + base_1 \rightarrow acid_2 + base_2$ $acid_1$ and $base_2$ are called "a conjugate pair"	covalent bond forms between acid and base
	examples	$HCl + NaOH \rightarrow$ $NaCl + H_2O$ neutralization	$HSO_4^- + NH_3 \rightarrow$ $NH_4^+ + SO_4^{2-}$ proton transfer	$Fe^{3+} + 6CN^- \rightarrow$ $Fe(CN)_6^{3-}$ complexation

Strong and Weak Acids and Bases

In Arrhenius acid-base systems, the terms strong and weak refer to the extent of ion formation. Strong acids are nearly 100% ionized; weak acids can often be much less than 1% ionized. The graph in Figure 9.2 below shows how percent ionization decreases as concentration increases.

Similarly, strong bases are 100% dissociated with weak bases undergoing much less dissociation.

Common strong acids: H_2SO_4, HCl, HNO_3, HBr, HI, $HClO_4$

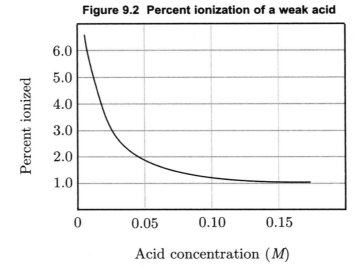

Figure 9.2 Percent ionization of a weak acid

Common strong bases: KOH, NaOH and other hydroxides of Group 1 metals, $Sr(OH)_2$, $Ba(OH)_2$

In Bronsted-Lowry systems, strong acids donate protons readily because the bond between hydrogen and the remainder of the molecule or ion is relatively weak. Strong bases form strong bonds between the donated proton and its new destination. Tables of Bronsted-Lowry acids and their conjugate bases as shown in Figure 9.3 are often presented in order of decreasing acid strength. Note that the conjugate of any acid is the corresponding base which you can determine by removing one H^+ from the acid form. It is often useful to think of a Bronsted-Lowry acid-base reaction as simply competition between two bases for a proton: the stronger base wins.

Figure 9.3

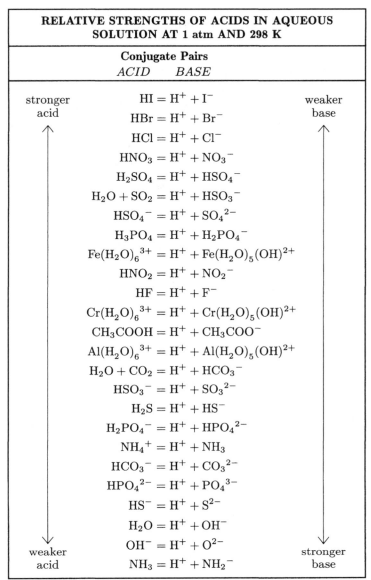

RELATIVE STRENGTHS OF ACIDS IN AQUEOUS SOLUTION AT 1 atm AND 298 K

Conjugate Pairs
ACID   BASE

stronger acid

$HI = H^+ + I^-$
$HBr = H^+ + Br^-$
$HCl = H^+ + Cl^-$
$HNO_3 = H^+ + NO_3^-$
$H_2SO_4 = H^+ + HSO_4^-$
$H_2O + SO_2 = H^+ + HSO_3^-$
$HSO_4^- = H^+ + SO_4^{2-}$
$H_3PO_4 = H^+ + H_2PO_4^-$
$Fe(H_2O)_6^{3+} = H^+ + Fe(H_2O)_5(OH)^{2+}$
$HNO_2 = H^+ + NO_2^-$
$HF = H^+ + F^-$
$Cr(H_2O)_6^{3+} = H^+ + Cr(H_2O)_5(OH)^{2+}$
$CH_3COOH = H^+ + CH_3COO^-$
$Al(H_2O)_6^{3+} = H^+ + Al(H_2O)_5(OH)^{2+}$
$H_2O + CO_2 = H^+ + HCO_3^-$
$HSO_3^- = H^+ + SO_3^{2-}$
$H_2S = H^+ + HS^-$
$H_2PO_4^- = H^+ + HPO_4^{2-}$
$NH_4^+ = H^+ + NH_3$
$HCO_3^- = H^+ + CO_3^{2-}$
$HPO_4^{2-} = H^+ + PO_4^{3-}$
$HS^- = H^+ + S^{2-}$
$H_2O = H^+ + OH^-$
$OH^- = H^+ + O^{2-}$
$NH_3 = H^+ + NH_2^-$

weaker acid

weaker base

stronger base

Neutralization

When an Arrhenius acid is mixed with an Arrhenius base, neutralization occurs. The quantitative principle involved specifies that one mole of mole H^+ exactly neutralizes one mole of OH^-. To manage the quantitative relationships in a neutralization reaction, there are generally four facts to be obtained: three are usually observations from the laboratory procedure under consideration; the fourth fact is calculated using these observations. For example, if the results of an experiment shows that 36.5 mL of 0.152 M $KOH_{(aq)}$ is neutralized by 27.3 mL of $HNO_{3(aq)}$, the apparent concentration of the HNO_3 solution can be calculated.

$$\frac{0.0365 \text{ L KOH}_{(aq)}}{0.0273 \text{ L HNO}_{3(aq)}} \times \frac{0.152 \text{ mol KOH}}{1.00 \text{ L KOH}_{(aq)}} = \frac{0.203 \text{ mol HNO}_3}{1.00 \text{ L HNO}_{3(aq)}}$$

Titration

When titration procedures are used to determine unknown concentration of an acid or base, there are two common methods of determining the point of exact neutralization.

Indicators In one kind of experiment, an indicator is used to indicate the endpoint of a titration, that is the conditions at which the exact neutralization has occurred and excess H^+ or OH^- ions become present in the solution. The choice of indicator is determined by the pH of the end product solution. Figure 9.4 shows the type of indicator needed, based on the properties of the acid/base reactants. Figure 9.5 is a list of frequently used indicators and the pH range over which their color changes are observed.

Figure 9.4 Titration: choosing an indicator

Neutralization of	Choose an indicator that changes color when the mixture of reactants is
Strong acid + strong base	neutral; pH approximately 7
Strong acid + weak base	is acidic; pH less than 7
Weak acid + strong base	is basic; pH greater than than 7

Figure 9.5 Common Acid-Base Indicators

Indicator	Approximate pH Range for Color Change	Color Change
methyl orange	3.2−4.4	red to yellow
bromcresol green	3.8−5.4	yellow to blue
litmus	5.5−8.2	red to blue
bromthymol blue	6.0−7.6	yellow to blue
thymol blue	8.0−9.6	yellow to blue
phenolphthalein	8.2−10	colorless to pink

pH Titrations In another kind of experiment, the pH of a reaction mixture is monitored as one reactant is added in small quantities – perhaps even drop-wise – to a solution of the other reactant. The observed values of pH are plotted as a dependent variable with volume of titrant added as the independent variable. Refer to your textbook for graphs of various pH titrations.

Hydrolysis of salts

Some salts dissolve in water to form solutions that are distinctly acidic or basic. Such salts are said to "hydrolyze". The conjugate bases of weak acids, often anions such as CH_3COO^- or CO_3^{2-}, react with water to form basic solutions.

$$\text{anion hydrolysis: } CO_3^{2-} + H_2O \rightleftharpoons HCO_3^- + OH^-$$

Some hydrated cations of transition elements such as Fe^{3+} react with water to form acidic solutions.

$$\text{cation hydrolysis: } Fe(H_2O)_6^{3+} + H_2O \rightleftharpoons Fe(H_2O)_5OH^{2+} + H_3O^+$$

Buffer solutions

A buffer solution is a solution of a weak acid and its conjugate base, at concentrations such that the pH of the solution does not change appreciably even though a substantial amount of H^+ or OH^- is added to the solution. The weak acid acts as a proton donor, neutralizing any added OH^-; the conjugate base accepts H^+, neutralizing any added acid.

Note that the capacity of a buffer is exceeded when the amount of proton donor added is greater than the amount of proton acceptor available in the buffer and vice versa.

Amphiprotic (amphoteric) compounds

In the Arrhenius system, an amphiprotic substance acts as an acid in the presence of a strong base and acts as a base in the presence of a strong acid. Aluminum hydroxide, $Al(OH)_3$, is a common example:

$$Al(OH)_3 + 3HCl \rightarrow AlCl_3 + 3H_2O$$

$$Al(OH)_3 + NaOH \rightarrow NaAlO_2 + 2H_2O$$

In the Bronsted Lowry system, an amphiprotic substance contains a proton that can be donated as well as an unshared electron pair that is available to accept a proton. The hydrogen carbonate ion, HCO_3^-, is a common example:

$$\text{as a proton donor: } HCO_3^- + OH^- \rightarrow H_2O + CO_3^{2-}$$

$$\text{as a proton acceptor: } HCO_3^- + H_3O^+ \rightarrow H_2O + H_2O + CO_2$$

from the TOPIC OUTLINE (website: apcentral.collegeboard.com)

III. Reactions

A. Reaction Types

    1. Acid-base reactions; concepts of Arrhenius, Bronsted-Lowry, and Lewis; coordination complexes; amphoterism

    2. Precipitation reactions

C. Equilibrium

    2 Quantitative treatment

        b. Equilibrium constants for reactions in solution

            (1) Constants for acids and bases; pK; pH

            (3) Common ion effect; buffers; hydrolysis

from the list of CHEMICAL CALCULATIONS

5. ... titration calculations

from the list of EQUATIONS & CONSTANTS

$$K_a = \frac{[H^+][A^-]}{[HA]}$$

$$K_b = \frac{[OH^-][HB^+]}{[B]}$$

$$K_w = [OH^+][H^+] = 10^{-14} \text{ @ } 25°C$$

$$= K_a \times K_b$$

$$pH = -\log[H^+], \ pOH = -\log[OH^-]$$

$$14 = pH + pOH$$

$$pH = pK_a + \log \frac{[A^-]}{[HA]}$$

$$pOH = pK_b + \log \frac{[HB^+]}{[B]}$$

$$pK_a = -\log K_a, \ pK_b = -\log K_b$$

from the list of RECOMMENDED EXPERIMENTS

6. Standardization of a solution using a primary standard

7. Determination of concentration by acid-base titration, including a weak acid or weak base

11. Determination of appropriate indicators for various acid-base titrations; pH determination

13. Determination of enthalpy change associated with a reaction

15. Synthesis of a coordination compound and its chemical analysis

19. Preparation and properties of buffer solutions

Multiple Choice Questions

Questions 1–5: The set of lettered choices below is a list of chemical compounds that are soluble in water and take part in acid-base reactions. Select the one lettered choice that best fits each numbered statement. A choice may be used once, more than once, or not at all.

(A) BF_3
(B) C_2H_5OH
(C) CH_3COOH
(D) $FeCl_3$
(E) $KHSO_3$

1. acts as a Lewis acid but not a Bronsted-Lowry acid

2. illustrates amphiprotic behavior in aqueous solution

3. dissolves as a weak electrolyte

4. undergoes hydrolysis in water to produce a solution that is acidic

5. reacts with an alcohol to form an ester

_____

6. Which is the best description of the behavior of a Lewis base in a chemical reaction?
   (A) electrophilic
   (B) heterophilic
   (C) homophilic
   (D) hydrophilic
   (E) nucleophilic

7. Which equation best illustrates the ionization of liquid ammonia?
   (A) $NH_3 \rightleftharpoons 3H^+ + N^{3-}$
   (B) $NH_3 \rightleftharpoons NH_2{}^- + H^+$
   (C) $NH_3 + NH_3 \rightleftharpoons NH_4{}^+ + NH_2^-$
   (D) $H_2O + NH_3 \rightleftharpoons H_3O^+ + NH_2^-$
   (E) $H_2O + NH_3 \rightleftharpoons NH_4{}^+ + OH^-$

8.   Which substance is a polyprotic acid in water solution?

I. CH_3COOH

II. C_2H_5OH

III. $H_2C_2O_4$

(A) I only
(B) III only
(C) II and III only
(D) I and II only
(E) I, II, and III

9.   Consider the three acids: HF, HSO_4^- and $H_2PO_4^-$.

Which list includes only conjugate bases of the acids given above?

(A) OH^-, HPO_4^{3-}
(B) F^-, SO_4^{2-} and HPO_4^{2-}
(C) OH^-, SO_4^{2-} and PO_4^{3-}
(D) OH^-, SO_4^{2-} and HPO_4^{2-}
(E) H_2F^+, H_2SO_4 and H_3PO_4

10. Which identifies the products that form when N_2O_5 dissolves in water?

(A) H^+ and NO_3^-
(B) NO_2 and H_2O_2
(C) H_2, O_2 and N_2
(D) NH_4^+, OH^- and O_2
(E) NH_4^+, NH_2^- and O_2

11. Which pair of equations illustrates the amphiprotic behavior of $H_2PO_4^-$?

I. $H_2PO_4^- + OH^- \rightarrow H_2O + HPO_4^{2-}$

II. $H_2PO_4^- + 2OH^- \rightarrow 2H_2O + PO_4^{3-}$

III. $H_2PO_4^- + H_3O^+ \rightarrow H_2O + H_3PO_4$

IV. $H_2PO_4^- + CH_3COOH \rightarrow CH_3COO^- + H_3PO_4$

V. $H_2PO_4^- + 2NH_2^- \rightarrow 2NH_3 + PO_4^{3-}$

(A) I and II only
(B) III and IV only
(C) II and V only
(D) I and V only
(E) IV and V only

12. Which occurs when potassium amide, KNH_2, dissolves in water?

    (A) K^+ is hydrolyzed to KH.
    (B) Nitrogen is oxidized from -3 to 0.
    (C) Nitrogen is reduced from $+1$ to 0.
    (D) A proton is transferred to NH_2^-.
    (E) A proton is transferred from NH_2^-.

13. Which oxide is a basic anhydride?

    I. Na_2O

    II. BaO

    III. ZnO

    (A) I only
    (B) III only
    (C) II and III only
    (D) I and II only
    (E) I, II, and III

14. Which range includes the pH that results when 0.10 mole $NaOH_{(s)}$ is added to 100. mL of 1.0 M HCl solution?

    (A) between 1 and 4
    (B) between 4 and 6.5
    (C) between 6.5 and 7.5
    (D) between 7.5 and 10
    (E) between 10 and 14

15. Each list contains at least one species that could illustrate amphiprotic behavior EXCEPT

    (A) HNO_3, HCl, HS^-
    (B) CO_3^{2-}, Br^-, NH_4^+
    (C) HCO_3^-, HSO_4^-, NH_3
    (D) $H_2PO_4^-$, NH_2^-, ClO_3^-
    (E) H_3PO_4, $Al(OH)_3$, $Zn(OH)_2$

16. All of the following apply to a solution prepared by adding potassium carbonate to water EXCEPT

    (A) The pH increases to a value greater than 7.
    (B) The aqueous phase of the system remains colorless.
    (C) The electric conductivity of the solution increases.
    (D) The concentration of OH^- ions in solution increases.
    (E) The dissolved species are molecules of potassium carbonate.

17. Which occurs when water is added to a solution of HCl?

	pH	**$[H_3O^+]$**
(A)	increases	decreases
(B)	increases	remains the same
(C)	decreases	increases
(D)	decreases	increases
(E)	remains the same	decreases

18. For CH_3COOH, $K_a = 1.8 \times 10^{-5}$

Which could be added to 1.0 liter of 0.10 M CH_3COOH to form a buffer solution with pH between 4 and 5?

    I.   0.050 mol HCl

    II.   0.050 mol NaOH

    III.   0.050 mol $NaCH_3COO$

(A) I only

(B) II only

(C) I and II only

(D) II and III

(E) I and III only

Questions 19-23: A sample containing precisely 25.0 mL of a solution of H_2SO_4 of unknown concentration is analyzed using a solution of NaOH of known concentration. The NaOH is titrated slowly into the H_2SO_4 solution from a buret. When neutralization occurs as shown by change in color of the indicator, the final value for the volume of $NaOH_{(aq)}$ is recorded.

NaOH solution	0.125 M
Starting volume, $NaOH_{(aq)}$	1.4 mL
Final volume, $NaOH_{(aq)}$	27.8 mL

19. Which identifies the spectator ions in this reaction?

(A) The only spectator ion is Na^+.

(B) The only spectator ion is SO_4^{2-}.

(C) Both Na^+ and SO_4^{2-} are spectator ions.

(D) Both Na^+ and HSO_4^- are spectator ions.

(E) There are no spectator ions in this reaction.

20. As this reaction proceeds, the concentration of Na^+ in the reaction mixture

(A) increases and the concentration of SO_4^{2-} increases

(B) increases and the concentration of SO_4^{2-} decreases

(C) remains the same and the concentration of SO_4^{2-} remains the same

(D) remains the same and the concentration of SO_4^{2-} increases

(E) remains the same and the concentration of SO_4^{2-} decreases

21. Which expression gives the molarity of the H_2SO_4 solution?

    (A) $\dfrac{0.0264 \times 0.125 \times 2}{0.0250}$

    (B) $\dfrac{0.0250 \times 2}{0.0264 \times 0.125}$

    (C) $\dfrac{0.0264 \times 0.125}{0.0250 \times 2}$

    (D) $\dfrac{0.0250 \times 0.125}{0.0264 \times 2}$

    (E) $\dfrac{0.0250 \times 0.125 \times 2}{0.0264}$

22. If, unknown to the technician, some water had been added to the unknown H_2SO_4 solution by mistake after its precise volume had been measured, which value in the table of observations, if any, would be changed? If there were a change, in which direction would the change occur?

    (A) The final volume of $NaOH_{(aq)}$ would be reported larger.

    (B) The final volume of $NaOH_{(aq)}$ would be reported smaller.

    (C) The molarity of the NaOH solution would be reported larger.

    (D) The molarity of the NaOH solution would be reported smaller.

    (E) No change would be reported in any of the three values.

23. If, unknown to the technician, some water had been added to the NaOH solution of known concentration by mistake before the final volume had been measured, which value in the table of observations, if any, would be changed? If there were a change, in which direction would the change occur?

    (A) The final volume of $NaOH_{(aq)}$ would be reported larger.

    (B) The final volume of $NaOH_{(aq)}$ would be reported smaller.

    (C) The molarity of the NaOH solution would be reported larger.

    (D) The molarity of the NaOH solution would be reported smaller.

    (E) No change would be reported in any of the three values.

24. Which gives the mass action expression for hydrolysis of the CO_3^{2-} ion?

    (A) $\dfrac{[H^+][OH^-]}{[CO_3^{2-}]}$

    (B) $\dfrac{[CO_3^{2-}]^2}{[H^+][OH^-]}$

    (C) $\dfrac{[HCO_3^-][H^+]}{[CO_3^{2-}]}$

    (D) $\dfrac{[CO_3^-][H^+]}{[HCO_3^-]}$

    (E) $\dfrac{[HCO_3^-][OH^-]}{[CO_3^{2-}]}$

25. When a sample of 0.0060 M NaOH is diluted with an equal amount of water, the pH of the resulting solution is closest to

    (A) 10.2
    (B) 10.8
    (C) 11.0
    (D) 11.5
    (E) 12.5

Free-Response Questions

26. A sample of 0.200 M solution of KHC_2O_4 (molar mass: 128 g) is used to determine the concentration of an unknown solution of KOH. Phenolphthalein is used as the indicator.

(A) What mass of KHC_2O_4 is required to prepare 750. mL of 0.200 M KHC_2O_4 solution?

(B) In one experiment, 25.0 mL of the unknown KOH solution is placed in a beaker. Added to the KOH solution in the beaker is 25.0 mL of water and a few drops of phenolphthalein. The KHC_2O_4 solution is titrated into the dilute base.

(1) The volume of KHC_2O_4 solution required to reach equivalence is 26.7 mL. What is the concentration of the original KOH solution?

(2) What specific observation identifies the endpoint of the titration?

(3) What is the molar concentration of $K^+_{(aq)}$ in the reaction mixture at the equivalence point?

(C) In a second experiment, 25.0 mL of the same unknown KOH solution as used in part (B) and 25.0 mL of water are placed in the beaker, as above. An additional 50.0 mL of water is added to the KOH solution in the beaker. Again the KHC_2O_4 solution of known concentration is titrated into the dilute base.

(1) Compared to the first experiment in (B) above, is the volume of $KHC_2O_{4(aq)}$ needed to reach the equivalence point of this second experiment greater, smaller or the same? Explain.

(2) Compared to the first experiment in (B) above, is the molar concentration of $K^+_{(aq)}$ in the reaction mixture at the equivalence point of this second experiment greater, smaller or the same? Explain.

(D) In another experiment, the technician, in error, used potassium acid oxalate hydrate, $KHC_2O_4 \cdot H_2O$ (molar mass: 146 g) without adjusting for different molar mass as if it were the assigned anhydrous compound for the preparation of the solution in Part (A). What effect would this have on the value subsequently reported for the molarity of the KOH solution compared to the true value for that molarity? Explain.

27. In the experiment described below, $Ba(OH)_{2(aq)}$ of unknown concentration is titrated against a precisely measured 25.0 mL volume of 0.0500 M $H_2SO_{4(aq)}$. Phenolphthalein, added to the acid solution, is used as an indicator. A system to measure conductivity is set up with a lamp connected in series with an ammeter and the solution to be tested. Assume constant temperature is maintained.

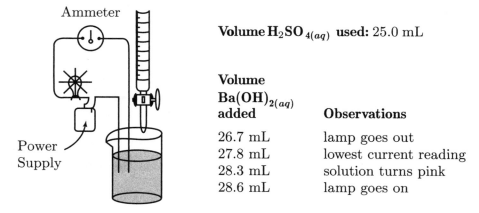

Volume $H_2SO_{4(aq)}$ used: 25.0 mL

Volume $Ba(OH)_{2(aq)}$ added	Observations
26.7 mL	lamp goes out
27.8 mL	lowest current reading
28.3 mL	solution turns pink
28.6 mL	lamp goes on

(A) Write the net ionic equation for this reaction.

(B) Explain why the light goes out but the ammeter reading never drops to zero.

(C) Explain why the light goes back on when the volume of $Ba(OH)_{2(aq)}$ added is 28.6 mL.

(D) Calculate the concentration of the barium hydroxide solution using the minimum value of conductance as the best indicator of the endpoint.

(E) A lower value for the concentration of the barium hydroxide would result if the color change of the indicator, instead of the minimum conductance reading, were used to indicate the endpoint. Explain.

28. In each of two flasks, a solution is prepared by placing 0.25 mol acetic acid, CH_3COOH, and 0.25 mol acetate ions, CH_3COO^-, in sufficient water to prepare 1.00 liter of solution. These two solutions are used for the experiments described in parts (C) and (D) below.

(A) What is a buffer solution? What characteristics make this solution a buffer solution?

(B) Draw the Lewis structures for the acetic acid molecule and the hydronium ion.

(C) Added to the solution in one flask is 0.15 mol $KHSO_{4(s)}$. What is the effect on $[CH_3COOH]$ and $[CH_3COO^-]$? Account for your answer in terms of numbers of moles of protons transferred. Has the capacity of the buffer been exceeded? Explain.

(D) Added to the solution in the second flask is 0.40 mol $NaOH_{(s)}$. What is the effect on $[CH_3COOH]$ and $[CH_3COO^-]$? Account for your answer in terms of numbers of moles of protons transferred. Has the capacity of the buffer been exceeded? Explain.

29. Write the formulas to show the reactants and the products for any FIVE of the laboratory situations described below. In all cases a reaction occurs. Assume that solutions are aqueous unless otherwise indicated. Represent substances in solution as ions if the substances are extensively ionized. Omit formulas for any ions or molecules that are unchanged by the reaction. You need not balance the equations.

(A) Excess 3 M ammonia is added to a solution of copper(II) sulfate.

(B) Excess sodium hydroxide solution is added to a suspension of aluminum hydroxide.

(C) Equal volumes of 0.10 M sodium hydrogen phosphate and 0.10 M hydrochloric acid are mixed.

(D) A lump of zinc metal is added to excess 1.0 M hydrochloric acid.

(E) Solid potassium oxide is added to water.

(F) Solid iron(II) sulfide is added to 6 M sulfuric acid.

(G) Solid ammonium carbonate is added to a warmed solution of sodium hydroxide.

(H) Sulfur dioxide gas is bubbled into water.

CHAPTER 10
ELECTROCHEMISTRY

Electrochemistry appears on the Advanced Placement Chemistry examination as net ionic equations, in galvanic and electrolytic cell problems, and in equilibrium and thermodynamics problems. You should be ready to address electrochemistry topics as part of many questions on the AP Chemistry exam.

Electrochemistry, that is, oxidation and reduction, deals with electrons moving in a path from the species oxidized to the species reduced. There are a number of mnemonic devices to help you remember this fundamental principle. One is OIL RIG, or Oxidation Is Loss while Reduction Is Gain of electrons.

OXIDATION NUMBERS

You will need to assign oxidation numbers to atoms in compounds or polyatomic ions as part of multiple choice questions as well as free response questions.

Rules for Oxidation Numbers

To facilitate the bookkeeping of electron transfers, oxidation numbers are assigned to each atom represented in each molecule or ion. The oxidation number should not be treated as the actual charge on an atom in the molecule or ion. Rules for assigning oxidation numbers (ON) can be summarized as follows:

1. Atoms in uncombined elemental form have zero as an oxidation number. Thus atoms in Fe, F_2, P_4 or S_8 all have ON of 0.

2. Monatomic ions have an oxidation number equal to the charge on the ion. Hence the ON of chloride is -1 while that of sulfide is -2.

3. When combined with other atoms, atoms of group 1 (alkali metals) always have ON of $+1$; atoms of group 2 (alkaline earth metals) always have ON of $+2$.

4. Combined oxygen always has ON of -2 (except for peroxides such as H_2O_2 where ON for oxygen is -1 or when combined with fluorine as OF_2 where ON for oxygen is $+2$).

5. Combined hydrogen always has ON of $+1$ (except for hydrides such as CaH_2 where ON for H is -1).

6. Combined fluorine always has ON of -1.

7. The ON of any other atom must be determined from its context. The sum of all ON within a compound must be zero (thus the ON of sulfur in H_2SO_4 must be $+6$). Within a polyatomic ion, the sum of ON must equal the overall charge on the ion. (Thus, the ON of phosphorus in $PO_4{}^{3-}$ must be $+5$.)

Oxidation = increase in ON, explained as loss of electrons

Reduction = decrease in ON, explained as gain of electrons

Balancing Redox Equations

There are various methods used to help with balancing redox equations. Here is one that never fails. It is sometimes called the "ion/water" or "ion/electron" method.

1. Starting with the given system or unbalanced equation, separate the reaction into two half-reactions. One must contain an oxidation while the other must include a reduction.

2. Balance each half-reaction separately.

   a. Balance the element oxidized or reduced.

   b. Balance any elements other than oxygen or hydrogen.

   c. Balance oxygen by adding water in the form of H_2O.

   d. In acidic solution, balance hydrogen by adding protons (H^+). In basic solution, balance hydrogen by adding water in the form H-OH, then immediately add the same number of hydroxides (OH^-) to the opposite side of the equation.

   e. Balance the total charge on each side of each half-reaction by adding electrons (e^-) to the side that is more positive.

3. Choose a multiplier for each half-reaction such that the number of electrons lost is equal to the number of electrons gained in the separate half-reactions. You will remember this technique from math class as being similar to determining the LCM (least common multiple). For example, if the oxidation loses 2 electrons while the reduction gains 3 electrons, the least common multiple is 6. Thus you should multiply the oxidation by 3 and the reduction by 2.

4. Add the two half-reactions together. The electrons should cancel out from each side of the equation. You may be able to cancel other species as well. Water as well as hydroxide and hydrogen ions will frequently cancel. Remember that H_2O and H–OH both mean water; they are written differently here to help balance oxygen and hydrogen efficiently.

5. Check to be sure that there is the same total charge on each side (It may be zero but is not required to be zero, just the same). Then check the total number of oxygen atoms on each side. If both the total charge and the number of oxygens balance, it is highly likely that you have balanced the overall reaction correctly.

   There are illustrations of balancing redox equations in the problems that follow this chapter.

Electrolytic and Galvanic Cells

In a galvanic (or voltaic) cell, chemical differences cause an electric difference to occur. Potential differences between substances cause electrons to flow as long as half-cells are separated by a salt bridge and external circuit. If the half-cells operate in physical contact with each other, the potential difference is presented as heat rather than as flow of electrons. No flow of electrons would be available to use in an external circuit.

An electrolytic cell represents the opposite situation: an electric current causes chemical changes to occur. An externally-supplied electric current forces chemical change to occur. In either electrolytic or galvanic cells, oxidation occurs at the anode and reduction at the cathode. Electrons flow through the external circuit from the anode to the cathode. Figure 10.1 provides a useful comparison of these two kinds of cells.

Figure 10.1 Comparing electrolytic and galvanic cells

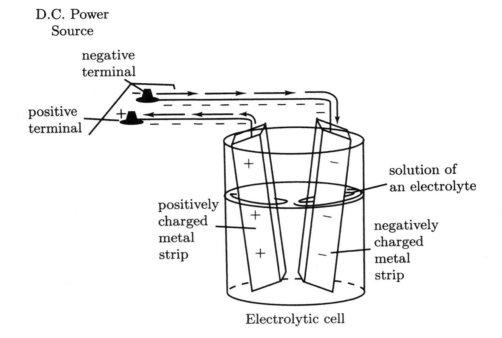

Electrolytic cell

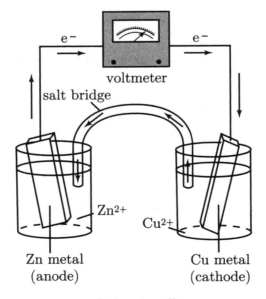

Galvanic cell

Using a table of standard reduction potentials

Since the oxidation and reduction half-reactions are combined to form an overall reaction to describe the electrochemical cell, so too are the values for the half-cells from the Standard Reduction Potential (SRP) table. To calculate $E°$, the standard voltage for a cell, you must combine the separate values for the two half-reactions. The SRP value for the reaction identified as oxidation (see #3 above) changes sign since the reduction equation from the table is reversed to become an oxidation. Thus $Zn^{2+} \rightarrow Zn$ shows a SRP of (-0.76 V); when expressed as an oxidation, $Zn \rightarrow Zn^{2+}$, the value for the half-reaction becomes ($+0.76$ V). The overall voltage for the cell described is determined by adding together the values for an oxidation and a reduction to give a total positive value. When the zinc oxidation half-reaction is combined with the reduction of copper(II) to copper, a total voltage of $+1.10$ V is created.

$$E° = E°_{oxidation} + E°_{reduction} = -(-0.76 \text{ V}) + (0.34 \text{ V}) = +1.10 \text{ V}$$

When a reaction or half-reaction is multiplied by any integer to balance an overall reaction, its standard reduction potential remains unchanged. (Three Cl_2 molecules pull six electrons with the same force that one Cl_2 pulls two electrons.)

Nonstandard conditions - The Nernst equation

The "nought" ("$°$") indicates standard conditions of pressure (one atmosphere), temperature (298 K), and concentration (1 M) of all solutions. If any of these conditions differs from the standard, then the voltage in the cell will differ. To calculate this different voltage, we can use the Nernst equation. In its full form, any difference in pressure, temperature, or concentration can be accommodated:

$$E = E° - \left(\frac{RT}{n\Im}\right) \ln Q$$

where $E° =$ the cell electromotive force (emf) as calculated from a table of standard reduction potentials (SRP)

$R = 8.314$ J mol^{-1} K^{-1}

$T =$ temperature in kelvins

$n =$ number of moles of electrons transferred in the redox

$\Im =$ Faraday's constant, 96,485 mol^{-1} V^{-1} (coulombs mol^{-1})

$Q =$ trial reaction quotient, including actual concentrations of all species.

However, because ion concentrations in solution are unaffected by the pressure over the surface of the solution and the standard temperature (298 K) is frequently assumed, a form of the Nernst equation is widely used to allow calculation with only differences in concentration. The constants of temperature, concentration, and faraday, in effect, combine to form a new constant. "Nernst lite" also uses base-10 log rather than the natural log.

$$E = E° - \left(\frac{0.0592}{n}\right) \log Q$$

VOCABULARY: OXIDATION/REDUCTION

Redox	a reaction process; oxidation accompanied by reduction
Oxidation	increase in oxidation number, explained as loss of electrons
Reduction	decrease in oxidation number, explained as gain of electrons
Reducing Agent	reactant species that loses electrons and gets oxidized; causes reduction of another species
Oxidizing Agent	reactant species that gains electrons and gets reduced; causes oxidation of another species
Oxidation Number	assigned to each atom in a chemical species to indicate apparent charge
Half-Reaction	a balanced equation that describes either oxidation or reduction; describes one half of the overall redox process

Electrochemical cells and their components

Electrochemical cell	a device in which the energy from a redox reaction is used to produce the flow of electrons; uses two separated redox half-reactions to generate electrical current through an external circuit; synonyms are voltaic, galvanic or simply chemical cell
Electrodes	surfaces of metal plates, grids, or wires, where electrons are gained and lost by metal atoms and ions; can also refer to the entire assembly of apparatus associated with the metal electrode
Anode	the electrode where oxidation occurs
Cathode	the electrode where reduction occurs
Cell potential	difference in electrical potential between the two electrodes of the two half cells, measured in volts
Reduction potential	conventional tabulated $E°$ values for half-reactions compared to the standard hydrogen electrode defined as $E° = 0.00$ V
Salt Bridge	a physical pathway that "completes the circuit" in some electrochemical cells; allows ions to migrate through a tube containing a solution of generally unreactive ions; this purpose can also be served by a porous barrier between half cells

Electrolytic cells and their components

Electrolytic Cell	a device that uses electrical energy to produce a chemical change that would otherwise not occur spontaneously
Electrolysis	the process of using electrical current to drive redox reactions
Electroplating	a process in which electrolysis used to deposit one metal on the surface of another metal or other substance.
Electrodes	are named anode and cathode using same definitions as with electrochemical cells, above

REDOX: REDUX – a brief summary

1. Oxidation and reduction always occur together since electrons must move from one species to another.

2. The number of electrons lost in oxidation must equal the number of electrons gained in reduction.

3. When comparing half-reactions to determine direction of electron flow, the half-reaction with the more negative reduction potential occurs at the anode of a galvanic cell as oxidation. (This species hold electrons more loosely.)

4. The half-reaction with the more positive reduction potential occurs at the cathode of a galvanic cell as reduction. (This species attracts electrons more strongly.)

5. When a reaction or half-reaction is multiplied by any integer to balance an overall reaction, its reduction potential remains unchanged. (Three Cl_2 molecules pull six electrons with the same force that one Cl_2 pulls two electrons.)

6. Commonly used line notation for a galvanic cell:

   anode/anode solution // cathode solution/cathode
   (where // represents a salt bridge or porous barrier)

   For example, in the Daniell cell, $Zn/Zn^{2+}//Cu^{2+}/Cu$ means

$$
\begin{array}{ll}
Zn \rightarrow Zn^{2+} + 2e^- & -(-0.76)\ V \\
\underline{Cu^{2+} + 2e^- \rightarrow Cu} & \underline{\quad 0.34\ V} \\
Cu^{2+} + Zn \rightarrow Zn^{2+} + Cu & E^\circ_{cell} = 1.10\ V
\end{array}
$$

7. A spontaneous reaction is signified by a positive cell voltage, a negative Gibbs Free Energy change, and an equilibrium constant with a positive exponent (constant > 1). To link these:

$$\Delta G^\circ = -n\Im E^\circ = -RT\ln K_{eq}$$

8. During electrolysis, the number of moles of electrons transferred depends on the current (amperage rate of flow) and the time during which the current flows. To link these ideas, remember that an ampere is defined as a coulomb per second:

$$A = \frac{C}{s} \qquad Ampere = \frac{coulomb}{sec}$$

The link between charge and moles of electrons (and indirectly the number of moles of reactants or products) is given by the definition of the faraday as 96,485 coulombs per mole of electrons:

$$\Im = \frac{96{,}485\ C}{mol\ e^-}$$

Thus, multiplying amperage by time by inverted faraday will reveal the number of moles of electrons released for an electrolytic reaction.

$$Mol\ e^-\ produced = amperage \times time \times \Im^{-1}$$

$$= \left(\frac{C}{sec}\right) \times (sec) \times \left(\frac{1\ mol\ e^-}{96{,}485\ C}\right)$$

from the TOPIC OUTLINE (website: apcentral.collegeboard.com)

III. Reactions

A. Reaction types

    3. Oxidation-reduction reactions

       a. Oxidation number

       b. The role of the electron in oxidation-reduction

       c. Electrochemistry: electrolytic and galvanic cells; Faraday's laws; standard half-cell potentials; Nernst equation; prediction of the direction of redox reactions

from the list of CHEMICAL CALCULATIONS

7. Faraday's laws of electrolysis (stoichiometry using moles of electrons)

9. Standard electrode potentials and their use; Nernst equation

from the list of EQUATIONS & CONSTANTS

$$I = \frac{q}{t} \text{ where I = current (amperes),}$$

$$q = \text{charge (coulombs) and } t = \text{time (seconds)}$$

$$\Delta G^\circ = -n\Im E^\circ = -RT \ln K_{eq}$$

$$\Delta G = \Delta G^\circ + RT \ln Q = \Delta G^\circ + 2.303RT \log Q$$

$$E_{cell} = E^\circ_{cell} - \frac{RT}{n\Im} \ln Q = E^\circ_{cell} - \frac{0.0592}{n} \log Q \text{ @ 25°C}$$

from the list of RECOMMENDED EXPERIMENTS

From the laboratory list:

 8. Determination of concentration by oxidation-reduction titration

 9. Determination of mass and mole relationship in a chemical reaction

20. Determination of electrochemical series

21. Measurements using electrochemical cells and electroplating

Multiple Choice Questions

Questions 1-5: The set of lettered choices is a list of ions that participate in oxidation reduction reactions. Select the one lettered choice that best fits each numbered description. A choice may be used once, more than once or not at all.

    (A) CO_3^{2-}

    (B) CrO_4^{2-}

    (C) MnO_4^-

    (D) NO_3^-

    (E) Fe^{2+}

1.   includes the species with a +7 oxidation number

2.   includes the species with a +6 oxidation number

3.   reacts with a reducing agent in acid solution such that the oxidation number of the metal becomes +2

4.   when oxidized, forms a cation with a +3 oxidation number

5.   reacts with a reducing agent in acid solution to form a brown gas

---

Questions 6-9:

$$Mg + 2Ag^+ \rightarrow Mg^{2+} + 2Ag \qquad E° = 3.17 \text{ V}$$

The above equation refers to a common electrochemical cell.

6.   Which expression gives the value of $\Delta G°$ in kJ mol^{-1} for this reaction?

    (A) $-2 \times 8.31 \times 3.17 \times 1000$

    (B) $\dfrac{(-2 \times 96,500 \times 3.17)}{8.31}$

    (C) $\dfrac{(-2 \times 96,500 \times 3.17)}{1000}$

    (D) $\dfrac{(-2 \times 96,500)}{(3.17 \times 8.31)}$

    (E) $\dfrac{(-2 \times 8.31 \times 3.17)}{1000}$

7.   When the equilibrium constant for this reaction is reported in scientific notation, the exponent for 10 is closest to

    (A) -100

    (B) -50

    (C)   0

    (D) $+50$

    (E) $+100$

8. Which set of non-standard concentrations does NOT produce a voltage for the cell different from its $E°$ value?

	$[Ag^+]$	$[Mg^{2+}]$
(A)	1.0	1.0
(B)	0.1	0.01
(C)	0.01	0.1
(D)	1.0	0.1
(E)	0.1	1.0

9. For the reaction above in one experiment, the observed value for E is +3.00 V. Which statement(s) apply to this cell?

    I.   Addition of more oxidizing agent will increase the value of E.

    II. Addition of products will decrease the value of $E°$.

    III. As this reaction proceeds, the value of E will decrease and the value of $E°$ will remain the same.

    (A) I only
    (B) II only
    (C) I and II only
    (D) I and III only
    (E) I, II, and III

Questions 10 and 11: Balance the following equation using redox techniques.

$$\dots H^+ + \dots Fe^{2+} + \dots MnO_4^- \rightarrow \dots Mn^{2+} + \dots Fe^{3+} + \dots H_2O$$

10. What is the sum of the coefficients in this balanced equation?

    (A) 6
    (B) 8
    (C) 16
    (D) 22
    (E) 24

11. What is the oxidizing agent in this reaction?

    (A) H^+
    (B) Fe^{2+}
    (C) MnO_4^-
    (D) Mn^{2+}
    (E) Fe^{3+}

Questions 12-15: The diagram below represents apparatus that can be used in the laboratory for the electrolysis of a 1.0 M solution of KI. A few drops of phenolphthalein solution are added. The terminals of the 12 volt DC power supply have been labeled with the appropriate charges. The current is supplied at the rate of 2.5 amperes.

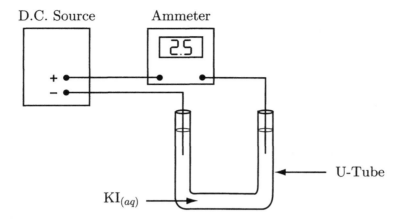

12. Which describes the appearance of the contents of the U-tube after the reaction has proceeded for a few minutes?

	anode chamber	cathode chamber
(A)	colorless	pink
(B)	brown	pink
(C)	brown	colorless
(D)	pink	colorless
(E)	colorless	colorless

13. Which describes the behavior of $K^+_{(aq)}$ ions in this system?

    (A) $K^+_{(aq)}$ ions migrate, as spectator ions, toward the cathode.

    (B) $K^+_{(aq)}$ ions migrate toward the anode where they become oxidized to $K_{(s)}$.

    (C) $K^+_{(aq)}$ ions migrate toward the anode where they become reduced to $K_{(s)}$.

    (D) $K^+_{(aq)}$ ions migrate toward the cathode where they become oxidized to $K_{(s)}$.

    (E) $K^+_{(aq)}$ ions migrate toward the cathode where they become reduced to $K_{(s)}$.

14. Which best describes the behavior in the chamber containing the electrode at which bubbles of gas are observed?

    (A) $O_{2(g)}$ is produced as the pH of the solution increases.

    (B) $O_{2(g)}$ is produced as the pH of the solution decreases.

    (C) $H_{2(g)}$ is produced as the pH of the solution increases.

    (D) $H_{2(g)}$ is produced as the pH of the solution decreases.

    (E) $O_{2(g)}$ is produced as the pH of the solution remains the same.

15. Which expression gives the time in hours needed to produce 0.10 mol $I_{2(aq)}$?

(A) $\dfrac{0.10 \times 2}{2.5 \times 60}$

(B) $\dfrac{0.10 \times 96{,}500}{2.5 \times 60 \times 60}$

(C) $\dfrac{0.10 \times 2}{2.5 \times 60 \times 96{,}500}$

(D) $\dfrac{0.10 \times 2 \times 96{,}500}{2.5 \times 60 \times 60}$

(E) $\dfrac{0.10 \times 2 \times 96{,}500 \times 60}{2.5}$

16. In the reaction:

$$2KMnO_4 + 3H_2SO_4 + 5H_2S \rightarrow 5S + 2MnSO_4 + K_2SO_4 + 8H_2O$$

the oxidation number of sulfur changes from

(A)   0 to -2
(B) $+5$ to -5
(C) -2 to 0
(D) -5 to $+5$
(E) $+6$ to $+4$

Questions 17-20: The diagram below represents a standard Fe^{2+}/Fe^{3+} half cell connected to a standard Pb^0/Pb^{2+} half cell using a salt bridge and an external circuit. The electrodes are numbered for purposes of identification.

$$Pb^{2+}{}_{(aq)} + 2e^- \rightarrow Pb^0{}_{(s)} \qquad E° = -0.13 \text{ volts}$$
$$Fe^{3+}{}_{(aq)} + e^- \rightarrow Fe^{2+}{}_{(aq)} \qquad E° = +0.77 \text{ volts}$$

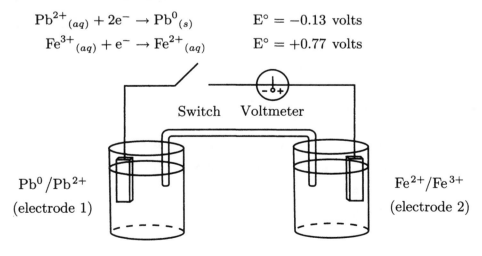

Switch   Voltmeter

Pb^0/Pb^{2+}
(electrode 1)

Fe^{2+}/Fe^{3+}
(electrode 2)

17. Which describes materials used for the construction of the standard Fe^{2+}/Fe^{3+} half cell?

> I.   The electrode is made of iron metal.
> II.   The source of Fe^{2+} could be $Fe(OH)_2$.
> III.   The source of Fe^{3+} could be $Fe(NO_3)_3$.

(A) I only
(B) III only
(C) II and III only
(D) I and III only
(E) I, II, and III

18. Which gives the changes in masses of the electrodes that occur during operation of the cell?

mass of electrode #1	mass of electrode #2
(A) decreases	increases
(B) decreases	remains the same
(C) remains the same	decreases
(D) increases	remains the same
(E) remains the same	increases

19. Which expression gives the value of the standard potential for the cell in volts?

(A) $0.77 - 0.13$
(B) $0.77 + 0.13$
(C) $0.77 + (2 \times 0.13)$
(D) $0.77 - (2 \times 0.13)$
(E) $0.77 + (2 \times (-0.13))$

20. Which gives the value for the potential of the cell in volts if $[Fe^{2+}]$ were changed from standard concentration to 0.010 M?

(A) $E° - 0.0592$
(B) $E° + 0.0592$
(C) $E° + \dfrac{0.0592}{2}$
(D) $E° - (2 \times 0.0592)$
(E) $E° + (2 \times 0.0592)$

21. The order of chemical activity of three metals is X > Y > Z. Which describes the behavior of these metals?

    I.   Atoms of X can reduce atoms of Z.
    II. Atoms of X can reduce cations of Z.
    III. Atoms of Y can reduce cations of X.

    (A) I only
    (B) II only
    (C) I and II only
    (D) II and III only
    (E) I, II, and III

22. Which gives the best description of ionic concentrations in the standard half-cell based on the half-reaction below?

    $$AgCl_{(s)} + e^- \rightarrow Ag_{(s)} + Cl^-_{(aq)} \qquad E° = 0.22 \text{ volts}$$

    (A) $[Ag^+] = [Cl^-] = 1.0\ M$
    (B) $[Ag^+] = [Cl^-] < 1.0\ M$
    (C) $[Ag^+] = [Cl^-] > 1.0\ M$
    (D) $[Cl^-] = 1.0\ M; [Ag^+] > [Cl^-]$
    (E) $[Cl^-] = 1.0\ M; [Ag^+] < [Cl^-]$

23. When the value of E° for a standard galvanic cell is greater than zero, which ranges apply to $\Delta G°$ and K_{eq} for the cell reaction?

	$\Delta G°$	K_{eq}
(A)	< 0	> 1
(B)	< 0	< 0
(C)	> 0	> 1
(D)	< 0	> 0 but < 1
(E)	> 0	> 0 but < 1

24. Which describes the oxidation number of nitrogen in each ion of NH_4NO_3?

    (A) Both nitrogen atoms have the same oxidation number with the same sign.
    (B) Both nitrogen atoms have the same oxidation number with the opposite sign.
    (C) The oxidation numbers of the two nitrogen atoms are +4 and +6, respectively.
    (D) The oxidation numbers of the two nitrogen atoms are −3 and +5, respectively.
    (E) The oxidation numbers of the two nitrogen atoms are −4 and +6, respectively.

25. A balanced half-reaction must illustrate conservation of

    I.   charge
    II. atoms
    III. molecules

    (A) I only
    (B) I and II only
    (C) II only
    (D) II and III only
    (E) I, II, and III

Free-Response Questions

26. A voltaic cell is set up prepared by connecting a standard Fe^{2+}/Fe^{3+} half cell to a standard Sn^0/Sn^{2+} half cell as shown below. Each half-cell contains 500 mL of 1.0 M solution.

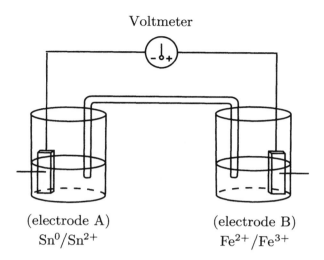

Voltmeter

(electrode A)   (electrode B)
Sn^0/Sn^{2+} Fe^{2+}/Fe^{3+}

(A) Write the overall net equation for this cell reaction and determine the initial cell voltage. Include the appropriate half-reactions and their electrode potentials.

(B) Identify the substance used as electrode A and the substance used as electrode B.

(C) Identify a substance suitable for use as a solute for the solution in the salt bridge. Explain the basis for your choice.

(D) In which direction do electrons flow in the external circuit? Explain.

(E) If 500 mL of water is added to the specified component(s) in the cell, what, if any, is the effect on the reduction potential, E, in each case below? Explain.

   (1) E of the Fe^{2+}/Fe^{3+} half cell, when 500 mL of water is added to that half-cell

   (2) E of the Sn^0/Sn^{2+} half cell, when 500 mL of water is added to that half-cell

   (3) overall E of the cell, when 500 mL of water is added to each half-cell

(F) What is the cell voltage when the system with the set of standard half-cells has operated long enough so that the concentration of Sn^{2+} has changed by 0.25 M?

27. The electrolytic cell shown below is used to plate silver metal onto a fork. Silver metal bars, solid silver nitrate and water are available for use in the cell.

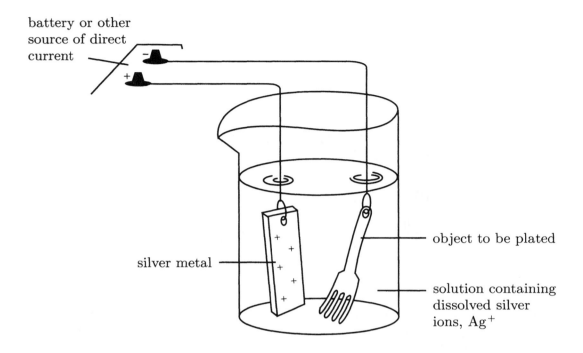

(A) Write the equation for the oxidation half-reaction that occurs in this system. What name is given to any electrode at which oxidation occurs?

(B) What mass of silver metal can be plated onto the spoon using a 30.0 amp current for 60.0 minutes?

(C) What effect is observed on the concentration of $Ag^+_{(aq)}$ in this system as the plating proceeds? Explain. (Assume excess mass for the silver metal bar.)

(D) A similar electrolytic device can be used to reduce $H^+_{(aq)}$ to $H_{2(g)}$. What volume of $H_{2(g)}$ measured at 298 K and 0.97 atm is produced if the number of faradays as used in part B is used in this similar device?

28. Answer the following questions about the analysis of iron-containing compounds using potassium permanganate solution.

(A) Write the balanced equation in ACID solution for the reaction below:

$$Fe^{2+} + MnO_4^- \rightarrow Fe^{3+} + Mn^{2+}$$

(B) To standardize a potassium permanganate solution, a 0.250 g sample of FAS (iron(II) ammonium sulfate hexahydrate; molar mass: 342 g) is dissolved in 25.00 mL distilled water, then acidified with sulfuric acid. The solution is then titrated with 35.00 mL potassium permanganate solution from a buret until a pale persistent purple color is attained. Calculate the molarity of the potassium permanganate solution.

(C) The standardized potassium permanganate solution is then used to titrate a solution made by dissolving a 0.500 gram sample of a mixture of iron(II) sulfate and sodium sulfate in 50.00 mL dilute sulfuric acid. A total of 10.21 mL of potassium permanganate solution is required to reach the pale purple endpoint. What is the mass percent of iron(II) sulfate in the original mixture?

(D) What would be the effect, if any, on the value for the reported standard molarity of potassium permanganate if the following errors were made? Explain each of your three answers.

(1) Some drops of water remained in the buret after cleaning but before the permanganate solution was added to the buret.

(2) The student neglected to run some permanganate solution through the tip of the buret before taking the initial reading.

(3) The student's lab partner spilled some FAS after weighing it but before titration.

CHAPTER 11

ORGANIC CHEMISTRY

Organic chemistry identifies the branch of chemistry that deals with most compounds of carbon. In AP chemistry, this refers to the study of several homologous series of hydrocarbons and their derivatives based on characteristic functional groups. The chemical and physical properties of these compounds as well as their nomenclature and molecular structure is included in this study.

You are expected to "know" a certain amount of information about organic chemistry. A good mental organization for this information can be based on this simple outline:

Compounds

- Hydrocarbons: names, formulas and properties

- Hydrocarbon derivatives: names, formulas and properties

- Structural formulas, Lewis structures, and isomers for the compounds specified above

Reactions

- Reactions from seven important categories

Much of the information "makes sense" based on molecular structure and related properties. Other information must simply be (ugh!) memorized.

HOMOLOGOUS SERIES OF HYDROCARBONS

You must be able to name and write molecular and structural formulas (Lewis structures) for a wide variety of hydrocarbons and their derivatives. Use this information to provide examples of substances, molecular structures and chemical reactions.

Alkanes

General formula: C_nH_{2n+2};

Name: prefix associated with number of carbon atoms as below plus the suffix -*ane*

Bonding: saturated: all carbon-carbon bonds are single bonds; sp^3 hybridization with tetrahedral geometry at each carbon atom

Typical reactions: (note: equations may not be balanced, intentionally)

$$\text{substitution:} \quad C_3H_8 + Cl_2 \rightarrow C_3H_7Cl + HCl$$

$$\text{combustion:} \quad C_4H_{10} + O_2 \rightarrow CO_2 + H_2O$$

Formula	Name	Structural Formula
CH_4	methane	$-\overset{\displaystyle |}{\underset{\displaystyle |}{C}}-$
C_2H_6	ethane	$-\overset{|}{\underset{|}{C}}-\overset{|}{\underset{|}{C}}-$
C_3H_8	propane	$-\overset{|}{\underset{|}{C}}-\overset{|}{\underset{|}{C}}-\overset{|}{\underset{|}{C}}-$
C_4H_{10}	butane	$-\overset{|}{\underset{|}{C}}-\overset{|}{\underset{|}{C}}-\overset{|}{\underset{|}{C}}-\overset{|}{\underset{|}{C}}-$
C_5H_{12}	pentane	$-C-C-C-C-C-$
C_6H_{14}	hexane	$-C-C-C-C-C-C-$
C_7H_{16}	heptane	$-C-C-C-C-C-C-C-$
C_8H_{18}	octane	$-C-C-C-C-C-C-C-C-$
C_9H_{20}	nonane	$-C-C-C-C-C-C-C-C-C-$
$C_{10}H_{22}$	decane	$-C-C-C-C-C-C-C-C-C-C-$

Alkenes

General formula: C_nH_{2n}

Name: prefix associated with number of carbon atoms as above plus the suffix -*ene*

Bonding: unsaturated: one carbon-carbon double bond with sp^2 hybridization (one *sigma* and one *pi* bond) and trigonal planar geometry at each "end" of the double bond; all other carbon-carbon bonds are single bonds.

Typical reaction

addition: $C_4H_8 + Br_2 \rightarrow C_4H_8Br_2$

Formula	Name	Structural formula
C_2H_4	ethene	$\rangle C = C \langle$
C_3H_6	propene	$\rangle C = C \langle$ with C
C_4H_8	butene	$-C-C=C\langle$ with C (2-butene)

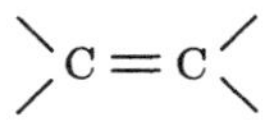

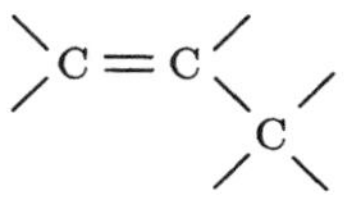

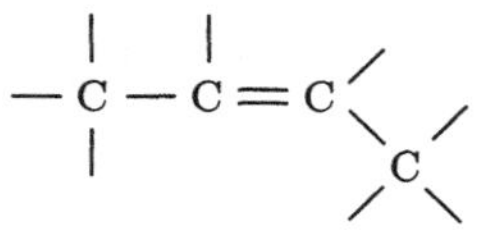

(Alkenes with more than four carbon atoms are rarely tested.)

Alkynes

General formula: C_nH_{2n-2}

Name: prefix associated with number of carbon atoms as below plus suffix -*yne*

Bonding: one carbon-carbon triple bond (*sp* hybridization and linear geometry at each "end" of the triple bond); all other carbon-carbon bonds are single bonds.

Typical reaction: addition: $C_4H_6 + Br_2 \rightarrow C_4H_6Br_4$

Formula	Name	Structural formula
C_2H_2	ethyne	$-C \equiv C-$
C_3H_4	propyne	$-C \equiv C-C-$
C_4H_6	butyne	$-C \equiv C-C-C-$ (1-butyne)

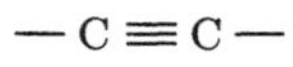

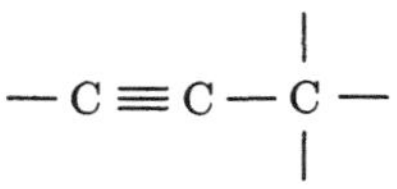

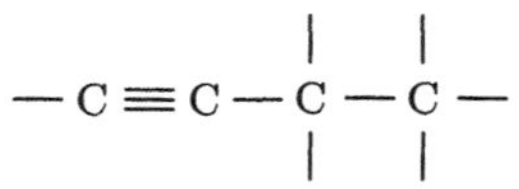

(Alkynes with more than four carbon atoms are rarely tested.)

Aromatic Hydrocarbons: the benzene series

General formula: C_nH_{2n-6}

Name: C_6H_6 - benzene; $C_6H_5CH_3$ - toluene;
$C_{10}H_8$ - naphthalene - actually two benzene "rings" sharing one side of their hexagons

Bonding: resonance structures; behaves more like a saturated compound than unsaturated; six carbon ring with **six sets of** sp^2 hybrid orbitals (six *sigma* bonds) and six delocalized (resonance) electrons equivalent to three *pi* bonds

Typical reactions: substitution: $C_6H_6 + Br_2 \rightarrow C_6H_5Br + HBr$

Formula	**Name**	**Structural Formula**
C_6H_5	benzene	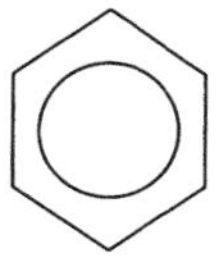
$C_6H_5CH_3$	toluene	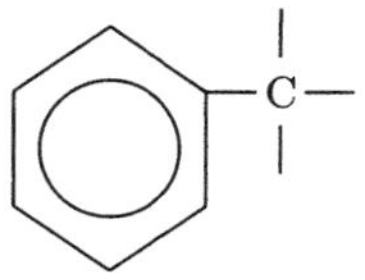
$C_{10}H_8$	naphthalene	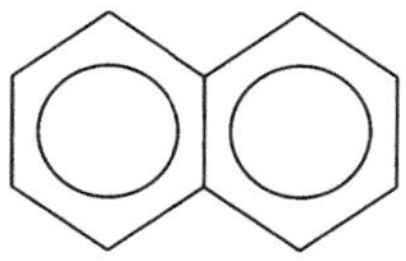

FUNCTIONAL GROUPS AND HYDROCARBON DERIVATIVES

Alcohols

primary alcohol

secondary alcohol

tertiary alcohol

$R\text{-}CH_2OH$	$R\text{-}CHOH\text{-}R'$	$R\text{-} R'\text{-} R''\text{-} OH$
1-butanol (primary)	2-butanol (secondary)	2-methyl-2-propanol (tertiary)

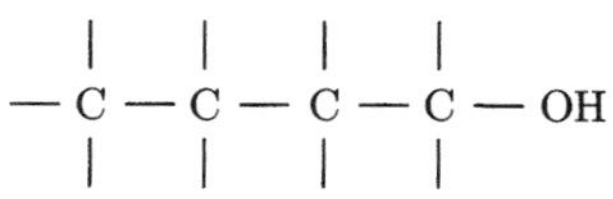

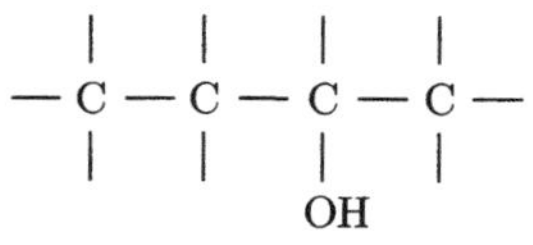

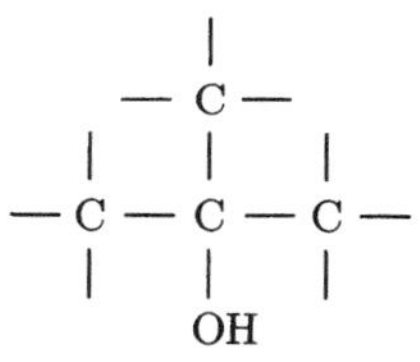

Ethers

$$R - O - R'$$

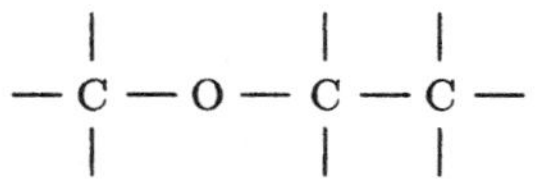

$$-\overset{\displaystyle |}{\underset{\displaystyle |}{C}} - O - \overset{\displaystyle |}{\underset{\displaystyle |}{C}} - \overset{\displaystyle |}{\underset{\displaystyle |}{C}} -$$

methyl ethyl ether

Aldehydes & Ketones

$$R - CHO \qquad\qquad R - CO - R'$$

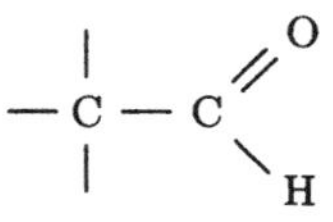

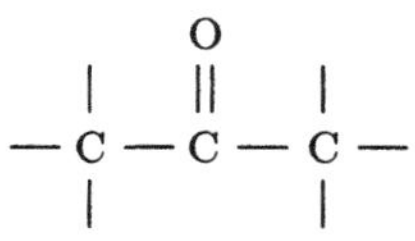

propanal   propanone

Carboxylic acids and esters

$$R - COOH \qquad\qquad R - COO - R'$$

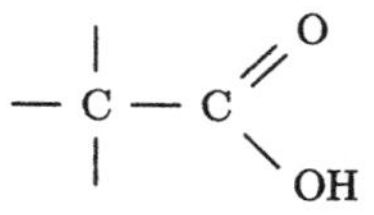

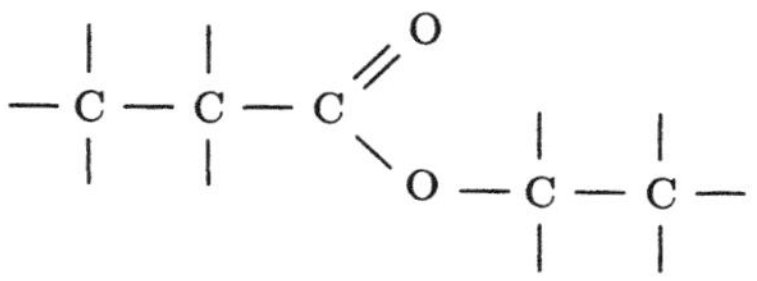

ethanoic acid   ethyl propanoate

Halohydrocarbons

$$R - X$$

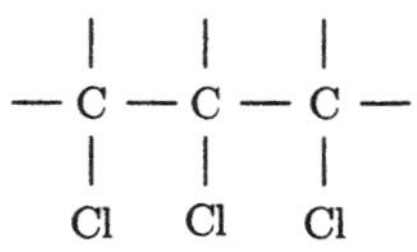

1,2,3-trichloropropane 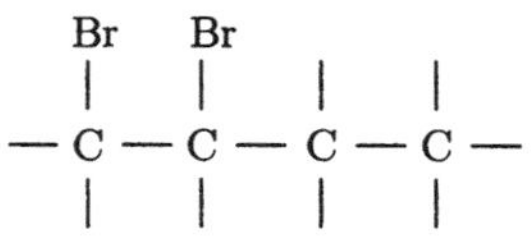   1,2-dibromobutane

Amines

$$R - NH_2$$

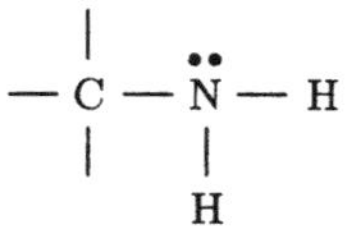

$$-\overset{\displaystyle |}{\underset{\displaystyle |}{C}} - \overset{\displaystyle \cdot\cdot}{\underset{\displaystyle |}{N}} - H$$
$$H$$

methyl amine

ISOMERISM: STRUCTURAL, *CIS-TRANS*, OPTICAL

You need to become familiar with simple illustrations of isomerism. Two compounds are isomers of each other when they have the same molecular formula but different structural formulas. Sometimes recognition and naming is sufficient. At other times, you will need to draw Lewis structures as well as assign names. Since isomers have different structures, they are likely to have different chemical and physical properties.

Structural isomerism:

butane   2-methylpropane

Cis-trans isomerism: $C_2H_2Cl_2$

cis-1,2-dichloroethene   *trans*-1,2-dichloroethene

See also 1,1-dichloroethene, below, a structural isomer.

Optical isomerism: (enantiomers)

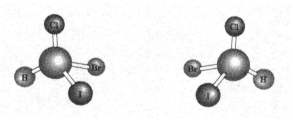

CHClBrI

two mirror images that cannot be superimposed

ORGANIC REACTIONS

Every reaction is simply a statement of a chemical property; that is, how one substance behaves in the presence of another when there is sufficient energy to start the reaction. Figure 11.1 provides a useful summary of some common organic reactions. Some organic reactions are slow starters with high energies of activation. For example, there is no reaction between gasoline and air (oxygen) until a spark is provided.

Compared to inorganic reactions in water solution, reactions between organic compounds generally proceed at much lower rates. Unlike charged ions, these neutral molecules have no special attraction for each other.

Figure 11.1 Some common organic reactions

	Reactants	Principal organic product
Substitution	saturated hydrocarbon + halogen, X_2	halohydrocarbon + HX
Addition	unsaturated hydrocarbon + halogen, X_2	halohydrocarbon
Esterification	carboxylic acid + alcohol	ester + HOH
Combustion	hydrocarbon + oxygen	CO_2 (or CO) + HOH
Oxidation	alcohol + oxidizing agent	aldehyde, ketone or acid
Polymerization **Addition Polymerization**	small unsaturated monomers	large saturated polymer
Condensation Polymerization	small monomers	large polymer and water or similar small molecules

from the TOPIC OUTLINE (website: apcentral.collegeboard.com)

IV. Descriptive Chemistry

   3. introduction to organic chemistry: hydrocarbons and functional groups
      (structure, nomenclature, chemical properties)

see also the section on Chemical Bonding (I.B.3): structural isomerism of simple organic molecules

from the list of CHEMICAL CALCULATIONS

No special calculations are associated with organic chemistry.

from the list of EQUATIONS & CONSTANTS

No entries are related specifically to organic chemistry.

from the list of RECOMMENDED EXPERIMENTS

22. Synthesis, purification and analysis of an organic compound

Multiple Choice Questions

Questions 1-5: The set of lettered choices is a list of functional groups found in organic compounds. Questions numbered 1–5 identifies some categories of organic compounds. For each numbered category, choose the functional group from the lettered list with which it is most closely associated.

(A) 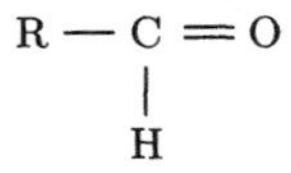 $R - C = O$
           $|$
           H

(B) $R - C = O$
           $|$
           $O - H$

(C) $R - C = O$
           $|$
           R'

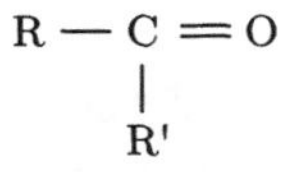

(D) 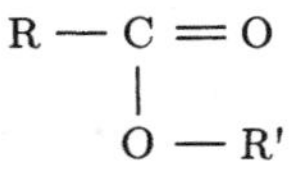 $R - C = O$
           $|$
           $O - R'$

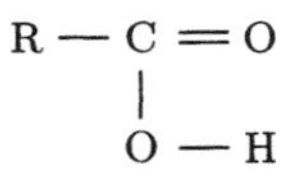

(E) $R - OH$

1.   alcohol

2.   carboxylic acid

3.   ketone

4.   ester

5.   aldehyde

6. Each of the following illustrates a pair of isomers EXCEPT

(A)

```
    H   H   H                          H   H   H
    |   |   |                          |   |   |
H — C — C — C — Cl      and     H — C — C — C — Br
    |   |   |                          |   |   |
    H   H   H                          H   H   H
```

(B)

```
    H   H   H                          H   H   H
    |   |   |                          |   |   |
H — C — C — C — OH      and     H — C — C — C — H
    |   |   |                          |   |   |
    H   H   H                          H   OH  H
```

(C)

```
    H   H   H                          H       H   H
    |   |   |                          |       |   |
H — C — C — C — OH      and     H — C — O — C — C — H
    |   |   |                          |       |   |
    H   H   H                          H       H   H
```

(D)

```
    H   H   H   H                          H       H
    |   |   |   |                          |       |
H — C — C = C — C — H      and     H — C — C = C — H
    |           |                          |
    H           H                          H
                                           |
                                       H — C — H
                                           |
                                           H
```

(E)

```
    H        Br                     H        H
     \      /                        \      /
      C = C          and             C = C
     /      \                        /      \
   Br        H                     Br        Br
```

7. Which organic compound is a weak electrolyte?

(A) C_2H_6

(B) CH_3Cl

(C) HCOOH

(D) CH_3OH

(E) $C_6H_{12}O_6$

8. In the fermentation of plant materials, which kind of compound is converted to alcohol?

(A) acid

(B) enzyme

(C) carbonate

(D) carbohydrate

(E) hydrocarbon

9. Which is the best description of the bonding associated with carbon atoms in most organic compounds?

    (A) formation of negative carbanions
    (B) formation of positive carbonium ions
    (C) formation of carbon free radicals with seven electrons
    (D) contribution of four electrons to the sharing of four pairs of electrons
    (E) transfer of electrons from carbon to elements with higher electronegativity

10. Which alcohol reacts with ethanoic acid to produce a four-carbon ester?

    (A) butanol
    (B) ethanol
    (C) methanol
    (D) pentanol
    (E) propanol

11. What is the total number of electron pairs shared between carbon atoms in a molecule of propene?

    (A) two
    (B) three
    (C) four
    (D) five
    (E) none

12. Which of the five compounds listed below has a molar mass that is different from the other four?

    (A) 2-butanol
    (B) 2-propanol
    (C) diethyl ether
    (D) 2-methyl-2-propanol
    (E) 2-methyl-1-propanol

13. Which of the following compounds represents the greatest degree of oxidation?

    (A) organic acid
    (B) ketone
    (C) alcohol
    (D) aldehyde
    (E) hydrocarbon

14. Which is an isomer of diethyl ether, $C_2H_5OC_2H_5$?

(A) C_4H_9OH

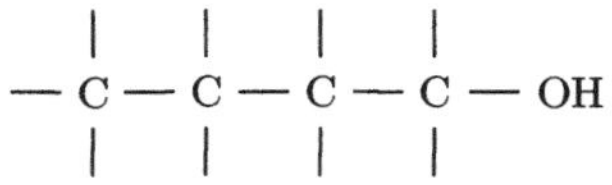

$$-\overset{|}{\underset{|}{C}}-\overset{|}{\underset{|}{C}}-\overset{|}{\underset{|}{C}}-\overset{|}{\underset{|}{C}}-OH$$

(B) $C_2H_5OCH_3$

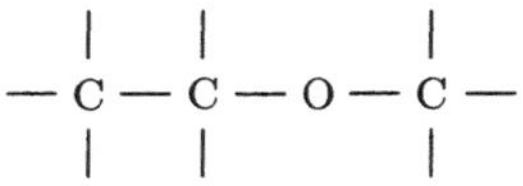

$$-\overset{|}{\underset{|}{C}}-\overset{|}{\underset{|}{C}}-O-\overset{|}{\underset{|}{C}}-$$

(C) C_2H_5OH

$$-\overset{|}{\underset{|}{C}}-\overset{|}{\underset{|}{C}}-OH$$

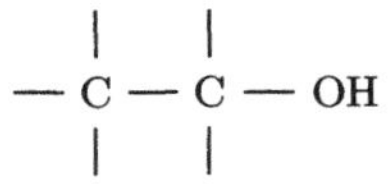

(D) $C_2H_5COC_2H_5$

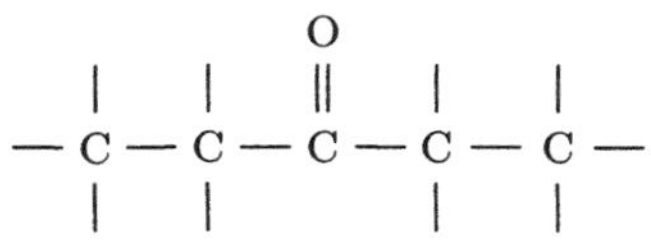

$$-\overset{|}{\underset{|}{C}}-\overset{|}{\underset{|}{C}}-\overset{O}{\overset{\|}{C}}-\overset{|}{\underset{|}{C}}-\overset{|}{\underset{|}{C}}-$$

(E) $C_2H_5SC_2H_5$

$$-\overset{|}{\underset{|}{C}}-\overset{|}{\underset{|}{C}}-S-\overset{|}{\underset{|}{C}}-\overset{|}{\underset{|}{C}}-$$

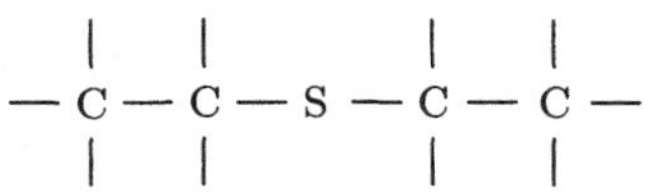

15. Which type of organic compound is most likely to participate in an addition reaction?

(A) acid
(B) alcohol
(C) aldehyde
(D) alkene
(E) ester

16. A molecule of ethene is similar to a molecule of methane in that they both have the same

(A) structural formula
(B) molecular formula
(C) number of carbon atoms
(D) number of carbon-carbon bonds.
(E) number of hydrogen atoms

17. The compound, 2-methylpentane, is an isomer of

(A) pentane
(B) 2-methylbutane
(C) 2,2-dimethylbutane
(D) 2,2-dimethylpentane
(E) 2,2-dimenthylpropane

18. When $KMnO_4$ reacts with ethanol in acid solution, manganese becomes

    (A) ionized
    (B) oxidized
    (C) reduced
    (D) catalyzed
    (E) dehydrated

19. Among the processes used to refine petroleum, which is most closely associated with differences in vapor pressure?

    (A) osmosis
    (B) filtration
    (C) distillation
    (D) crystallization
    (E) centrifugation

20. Which is a correct comparison of carbon-carbon single bonds to carbon-carbon double bonds in hydrocarbon molecules?

    I.   The bond length of the single bond is greater.
    II. The bond energy of the single bond is greater.
    III. The reactivity of the single bond is greater.

    (A) I only
    (B) III only
    (C) I and II only
    (D) II and III only
    (E) I, II, and III

21. Which formula represents an unsaturated hydrocarbon?

    (A) C_3H_6
    (B) C_3H_8
    (C) $C_2H_5CH_2OH$
    (D) C_2H_5CHO
    (E) C_2H_5COOH

22. A 1.0 molal aqueous solution of methanol differs from a 1.0 molal aqueous solution of acetic acid in that the aqueous solution of methanol

   (A) contains two phases
   (B) has a higher density
   (C) conducts electricity
   (D) turns blue litmus to red
   (E) contains only molecules

23. What is the total number of shared electron pairs between all adjacent carbon atoms in a propyne molecule?

   (A) 1
   (B) 2
   (C) 3
   (D) 4
   (E) 5

24. Which hybridization provides the best explanation for the geometry of the *sigma* (σ) bonds in formaldehyde, HCHO?

   (A) sp
   (B) sp^2
   (C) sp^3
   (D) dsp^3
   (E) d^2sp^3

25. Which general formula applies to hydrocarbons with one double covalent bond between adjacent carbon atoms?

   (A) C_nH_{2n+2}
   (B) C_nH_{2n}
   (C) C_nH_{2n-2}
   (D) C_nH_{2n-4}
   (E) C_nH_{2n-6}

Free-Response Questions

26. Use principles of chemical bonding to answer the following questions about several organic compounds.

    (A) Write the structural formulas and the IUPAC name for each of the two isomers of $C_2H_4Br_2$.

    (B) Which of the two isomers has the greater polarity? Explain.

    (C) Write labeled structural formulas to show the difference in bonding between methyl ethanoate and ethyl methanoate.

    (D) The value of K_a for the aromatic acid, phenol, C_6H_5OH, is 1.3×10^{-10}. The value of K_a for the carboxylic acid, ethanoic acid, CH_3COOH, is 1.8×10^{-5}. Using this information, discuss the difference in bond strength for the $O-H$ bond in each substance.

27. Consider the formulas of the organic compounds listed below

    C_2H_6 $C_2H_2Br_2$ C_2H_5OH CH_3COOH $CHClBrI$ C_4H_9OH

    (A) Choose the formula of the substance above whose molecules contain one carbon-carbon double bond. Write the names and structural formulas of its *cis* and *trans* isomers.

    (B) Choose the formulas of the two substances above that react to form an ester. Write the name and structural formula of that ester.

    (C) Choose the formula of the substance above that donates a proton most readily to OH^- in water solution. Write the name and structural formula of that proton donor.

    (D) Choose the formula of the substance above whose molecules exhibit optical isomerism. Write a pair of structural formulas to illustrate that isomerism.

    (E) Choose the formula of the alcohol that exists as four isomers. Name and draw the structural formula for each of the four isomers.

28. For each of the following, name the category of reaction that occurs and write the structural formula for the principal organic product.

    (A) C_4H_9OH is mixed with CH_3COOH in the presence of concentrated H_2SO_4.

    (B) $Br_{2(\ell)}$ is added to *cyclo*-C_6H_{10}.

    (C) A solution of $KMnO_4$ is added to an acidified solution of C_3H_7OH.

    (D) C_2H_4 is heated in the presence of a catalyst.

CHAPTER 12
DESCRIPTIVE CHEMISTRY

"Descriptive chemistry" is a nebulous term that might be interpreted to include any chemistry that is non-quantitative. Much of "descriptive chemistry" addresses the physical and chemical properties of various elements and compounds. Those properties include the interactions of these various elements and compounds with each other. Other information may also be included in "descriptive chemistry". For example, nuclear chemistry might be considered a descriptive topic within the AP Chemistry arena. You may be asked to write and balance nuclear equations and to describe how nuclear changes proceed. In addition, half-life problems from reaction kinetics are frequently presented in nuclear chemistry terms. Much of what you are expected to know can be gleaned from your first-hand experiences in the laboratory. How better to learn that copper(II) ion is blue in aqueous solution or that zinc metal reacts vigorously with strong acid to produce hydrogen gas? Be absolutely sure to review your laboratory notebook before taking the AP Chemistry examination!

WRITING NET IONIC EQUATIONS: ALWAYS ELIMINATE SPECTATOR IONS!

One question in the free response section of the AP Chemistry examination always requires you to write net ionic equations to demonstrate your knowledge of chemical changes. You have already addressed this topic thoroughly in the introductory section of this book. The products of many of these reactions can be predicted by a qualitative background in descriptive chemistry. In your reaction, be sure to show the formula for any species that is present

- as a dissolved ion that undergoes chemical change
- in solid, liquid or gaseous phase
- as a molecular species

Spectator ions (those ions that exist in the same form on both reactant and product side of a complete ionic equation) must be eliminated since they have not participated in any chemical change.

Periodic Table (See also Chapter 1.)

Be sure that you can identify and locate the representative (main group) elements, active metals, transition metals, inner transition metals (rare earth elements), metalloids, non-metals, and noble gases. You should be able to write convincingly about their chemical behavior and electron configuration as well as the relationship between those two characteristics. Periodic trends in several properties are frequently addressed on the AP Exam. The summary below is a reminder of the principles discussed in Chapter 1.

Periodic Trends

Within a period, as atomic number increases:

- atomic radius decreases
- first ionization energy increases
- electron affinity becomes more negative
- ionic radius for isoelectronic species decreases
- melting point decreases

Within a group, as atomic number increases:

- atomic radius increases
- first ionization energy decreases
- ionic radius increases
- melting point increases

ACIDS, BASES, AND SALTS (See also Chapter 9.)

You should know the names and formulas of the common acids that are classified as strong because they dissociate completely in solution (HCl, HBr, HI, $HClO_4$, HNO_3, H_2SO_4). All other acids (species that can donate a proton in water solution) are considered to be weak. The relative strengths of these weak acids can be compared by referring to their K_a values. When you compare the strength of oxyacids that are identical except for the number of oxygens included, the molecule with more oxygen atoms is the stronger acid ($HNO_3 > HNO_2$; $H_2SO_4 > H_2SO_3$). If oxyacids are similar except for the central non-metal atom, the acid with the non-metal of greater electronegativity is the stronger acid ($H_2SO_4 > H_2SeO_4$).

The strong bases are the hydroxides of all Group 1 metals and the heavier Group 2 metals such as $Ba(OH)_2$ and $Sr(OH)_2$; some authorities include $Ca(OH)_2$. All other hydroxides and proton acceptors are considered to be weak bases.

Acidic and basic anhydrides are oxides that react with water to form easily recognized acids or bases. Oxides of non-metals are acid anhydrides (SO_2 forms H_2SO_3; SO_3 forms H_2SO_4). Oxides of metals are basic anhydrides (Li_2O forms $LiOH$; BaO forms $Ba(OH)_2$). Note that there is no change in the oxidation number of the non-metal or metal element when the oxide dissolves in water to form the acid or the base. When anhydrides react with water, there is no oxidation or reduction. An acid anhydride combines with a basic anhydride to form a salt ($CaO + SO_3 \rightarrow CaSO_4$). Note that even though there is no water is in sight, the reaction can still be considered an acid-base reaction!

Solutions of salts in water: hydrolysis

Recall that some salts dissolve in water to form acidic or basic solutions. See Chapter 9 for the discussion of hydrolysis. The information below can be used to predict the acidity of most salt solutions:

Salts of strong acid & weak base $\rightarrow$ acidic solution;
eg. $FeCl_3$ - a salt derived from HCl, a strong acid, and $Fe(OH)_3$, a weak base

Salts of weak acid & strong base $\rightarrow$ basic solution;
eg. K_2CO_3 - a salt derived from H_2CO_3, a weak acid, and KOH, a strong base

Salts of strong acid & strong base $\rightarrow$ neutral solution;
eg. Na_2SO_4 - a salt derived from H_2SO_4, a strong acid, and NaOH, a strong base

Salts of weak acid & weak base $\rightarrow$ nature of solution not easily predicted

Colors of solutions

You may be asked to discuss the presence or absence of certain ions in solution based on information about color. See Figure 12.1 for a list of ions often used in AP Chemistry.

Figure 12.1 Colors of ions in aqueous solution

ion	color
Cu^{2+}	blue
Ni^{2+}	green
CrO_4^{2-}	yellow
$Cr_2O_7^{2-}$	orange
MnO_4^{-}	violet
Co^{2+}	pink
Fe^{3+}	pale yellow

In general, the presence of unpaired d-sublevel electrons is associated with colored precipitates/solutions/or gases. Otherwise, products are likely to be white or colorless in solution.

OXIDIZING AND REDUCING AGENTS (See also Chapter 10.)

Be able to identify common oxidizing and reducing agents in order to discuss oxidation-reductions reactions. See Figure 12.2 (below). You should always look for species that contain atoms that have been pushed to an extreme in either accepting or donating electrons. For example, the oxidation number of the manganese atom in permanganate MnO_4^- is +7. It has already "lost" all seven $4s$ and $3d$ electrons that it was assigned as a free atom; it has no more valence electrons to lose. Permanganate can only be reduced by gaining electrons and therefore is a vigorous oxidizing agent. For similar reasons, chromate and dichromate, where ON for chromium is +6, are also excellent oxidizing agents.

Species that are in the middle of their possible range of oxidation numbers can be either oxidized or reduced, depending on the context. Thus hypochlorite, ClO^-, can be reduced to chlorine, Cl_2, or oxidized to chlorite, ClO_2^-, depending on the relative strength of its redox partner.

Figure 12.2 Commonly used oxidizing and reducing agents

Common oxidizing agents and their reduction products:

$$MnO_4^- \rightarrow Mn^{2+} \text{ (acidic); } MnO_2 \text{, occasionally } MnO_4^{2-} \text{ (basic)}$$

$$NO_3^- \rightarrow NO_2, NO \text{ or } NH_4^+ \text{ (acidic)}$$

$$H_2O_2 \rightarrow H_2O \text{ (acidic)}$$

$$H_2SO_4 \rightarrow SO_2 \text{ (acidic)}$$

Active nonmetals $\rightarrow$ anions (such as $Br_2 \rightarrow Br^-$)

Common reducing agents and their oxidation products:

$$SO_2 \rightarrow SO_4^{2-}$$

$$SO_3^{2-} \rightarrow SO_4^{2-}$$

$$Sn^{2+} \rightarrow Sn^{4+}$$

Active metals $\rightarrow$ cations (such as $Ba \rightarrow Ba^{2+}$)

ORGANIC CHEMISTRY (See also Chapter 11.)

Organic chemistry is the study of most of the compounds that contain carbon atoms. Molecules that contain only carbon atoms and hydrogen atoms are called hydrocarbons. You should know how to identify and name three series of hydrocarbons, the alkanes, alkenes, and alkynes and their derivatives (alcohols, ethers, and so on). Similarly, you should also be familiar with the properties of benzene, C_6H_6, and toluene, $C_6H_5CH_3$. It is also useful to know a handful of important reactions. See Chapter 11 for a review of this information. Physical and chemical properties of simple organic compounds should be included as exemplary material for the study of other areas such as bonding, equilibria involving weak acids and bases, kinetics, colligative properties, and stoichiometric determinations of empirical and molecular formulas. It is not necessary to conduct an in-depth survey of organic chemistry at this point since a successful performance on the AP Chemistry examination tends to point you towards a full-fledged course in organic chemistry anyway!

COORDINATION COMPOUNDS

Although coordination chemistry is a fascinating topic in its own right, it is not a major topic of the AP Chemistry course. The bonding is interesting as ligands donate a pair of electrons from a filled orbital of the ligand to an empty orbital in the central metal cation to form a coordinate covalent bond.   You should be able to name coordination compounds by identifying and naming the complex ion, then pairing it with another ion to complete the compound.   You should also be familiar with the geometry of shared electron pairs (see Chapter 2) and with isomer formation based on location of ligands. These compounds may show spectacular colors so it is well worth your time in lab to synthesize such a compound. A list of common ligands is given in Fig 12.3.

Figure 12.3 Common ligands and nomenclature

Ligand	Name
NH_3	ammine- (note spelling, two "m"s)
CN^-	cyano-
H_2O	aqua- (occasionally, aquo-)
OH^-	hydroxy-
Cl^-	chloro-
$C_2O_4{}^{2-}$	oxalato- (a bidentate ligand - two electron pairs available for sharing)

Naming Coordination Compounds

1.   In a compound, the cation is named first, then the anion.

2.   In a complex ion, the ligands are named first, then the metal cation.

3.   Ligands are named by adding -o to the anion name (chloride becomes chloro-, cyanide becomes cyano-, and so on).

4.   If there is more than one of a given ligand, the number is denoted by a Greek counting prefix (di-, tri-, tetra-, penta-, etc.).   If the name of the ligand already contains a Greek counting prefix, then the prefixes bis-, tris-, tetrakis-, pentakis-, and so on are used instead.

5.   The oxidation number of the metal cation is denoted by use of a Stock system Roman numeral in parentheses.

6.   If there is more than one type of ligand, they are listed in alphabetical order.

7.   If the complex ion is an anion, its name must end in -ate.

Examples:

$[Co(H_2O)_4Cl_2]Br$	tetraaquadichlorocobalt(III) bromide
$[Co(H_2O)_6]^{3+}$	hexaaquacobalt(III) cation
$[CoCl_4]^-$	tetrachlorocobaltate(III) anion

NUCLEAR CHEMISTRY

You should be able to write and balance nuclear equations that include alpha, beta, and positron emission and electron absorption (K-capture). Figure 12.4 contains a list of frequently used symbols. As long as you remember to maintain conservation of charge (the total number of protons on each side of the equation must be the same) and conservation of mass (the total number of protons plus neutrons on each side of the equation must be the same), balancing nuclear equations is easy. The balanced equation below represents the bombardment of an isotope of aluminum with alpha particles.

$$_2^4\text{He} + {}_{13}^{27}\text{Al} \rightarrow {}_{15}^{30}\text{P} + {}_0^1\text{n}$$

Chemists frequently do not include gamma radiation in these equations since the high energy gamma rays have neither charge nor mass and therefore do not affect those values.

Radioactive decay is a first-order process. Half-life as discussed in Chapter 6 applies to the radioactive decay of most isotopes.

Figure 12.4

SYMBOLS USED IN NUCLEAR CHEMISTRY		
electron (beta particle)	$_{-1}^{0}\text{e}$	β^-
position	$_{+1}^{0}\text{e}$	β^+
proton	$_1^1\text{H}$	p
alpha particle	$_2^4\text{He}$	α
neutron	$_0^1\text{n}$	n
gamma radiation		γ

from the TOPIC OUTLINE (website: apcentral.collegeboard.com)

I.   Structure of Matter

    C. Nuclear chemistry: nuclear equations, half-lives and radioactivity; chemical application

IV. Descriptive Chemistry

    1. Chemical reactivity and products of chemical reactions

    2. Relationships in the Periodic Table: horizontal, vertical and diagonal with examples from alkali metals, alkaline earth metals, halogens, and the first series of transition elements.

from the list of CHEMICAL CALCULATIONS

No calculations are especially associated with Descriptive Chemistry.

from the list of EQUATIONS & CONSTANTS

Beer's law for colored solutions

$$A = abc \text{ where}$$

A = absorbance

a = molar absorptivity

b = path length

c = concentration (molarity)

from the list of RECOMMENDED EXPERIMENTS

14. Separation and qualitative analysis of cations and anions

15. Synthesis of a coordination compound and its chemical analysis

17. Colorimetric or spectrophotometric analysis (see Beer's Law)

18. Separation by chromatography

22. Synthesis, purification, and analysis of an organic compound

Multiple Choice Questions

Questions 1-5: The set of lettered choices below is a list of observations that refer to the numbered list of mixtures immediately following. For each numbered mixture, select the one lettered observation that is most closely associated with the mixture specified. A choice may be used once, more than once or not at all.

(A) forms a colorless solution with a colored precipitate
(B) forms a colorless solution with white precipitate
(C) forms a colored solution with no precipitate
(D) forms a colorless solution only with no precipitate or evolution of gas
(E) forms a solution with no precipitate, accompanied by evolution of gas

1. $CaCl_{2(aq)} + NaCl_{(aq)}$

2. $AgNO_{3(aq)} + KCl_{(aq)}$

3. $K_2CrO_{4(aq)} + $ excess $Ba(NO_3)_{2(aq)}$

4. dilute $HCl_{(aq)} + KHCO_{3(aq)}$

5. $CrCl_{3(s)} + $ dilute $HNO_{3(aq)}$

---

6. Which mixture of solids dissolves completely in dilute hydrochloric acid but not in distilled water?

    (A) $NaCl$ and KNO_3
    (B) $Ca(NO_3)_2$ and $NaCl$
    (C) Na_2CO_3 and $CaCl_2$
    (D) $NaHCO_3$ and $Ca(NO_3)_2$
    (E) $Ca(HCO_3)_2$ and KNO_3

7. Which occurs when CO_2 is bubbled into a solution of limewater, $Ca(OH)_2$?

    (A) A white precipitate forms that redissolves.
    (B) A light blue precipitate accumulates at the top of the mixture.
    (C) The liquid separates into two immiscible layers.
    (D) A noticeable evolution of energy occurs with a corresponding increase in temperature.
    (E) A noticeable absorption of energy occurs with a corresponding decrease in temperature.

Questions 8-12: The set of lettered choices below is a list of ions in separate water solutions that are to be tested with a solution of NaOH. Each of the numbered mixtures immediately following contains NaOH and only one species of ion from those listed. Select the one ion from the list of lettered choices that is most closely associated with the mixture specified. A choice may be used once, more than once or not at all.

(A) Al^{3+}

(B) $Cr_2O_7{}^{2-}$

(C) Mg^{2+}

(D) $NH_4{}^+$

(E) Ni^{2+}

8. When solution of NaOH is added, the color of the solution changes from orange to yellow.

9. When NaOH solution is added and the mixture warmed, the smell of ammonia is detected.

10. When excess NaOH solution is added, a white precipitate forms that redissolves.

11. When NaOH solution is added, a persistent white precipitate forms.

12. When NaOH solution is added, a persistent green precipitate forms.

---

13. Which pair of solutions, when mixed, produces a white precipitate?

(A) $AgNO_3 + NaCl$

(B) $AgNO_3 + K_2CrO_4$

(C) $AgNO_3 + KMnO_4$

(D) $Mn(NO_3)_2 + Na_2S$

(E) $AgNO_3 + Na_2S$

14. Each of the following cations forms a colored aqueous solution EXCEPT

(A) Ca^{2+}

(B) Ni^{2+}

(C) Cu^{2+}

(D) Mn^{2+}

(E) Co^{2+}

Questions 15-19:

    (A) Na
    (B) S
    (C) P
    (D) Si
    (E) Zn

The set of lettered choices above is a list of chemical elements. For each numbered statement below, select the element that is most closely associated with that statement. A lettered choice may be used once, more than once, or not at all.

15. One of its oxides is a gas that dissolves in water to form a strong acid.

16. Its oxide is a solid that dissolves in water to form a strong base.

17. Its oxide is a network solid that is insoluble in water.

18. Its oxide is a solid that dissolves in both excess acid and excess base.

19. One of its oxides is a white solid that dissolves in water to form a weak acid.

---

20. Which group of the Periodic Table includes elements with names meaning "the Sun element", "the new one", "the lazy one", "the hidden one", and "the strange one"?

    (A) alkali metals
    (B) alkaline earth metals
    (C) chalcogens
    (D) halogens
    (E) noble gases

21. Which gives the correct trend in atomic number and corresponding atomic radius within the third period of the Periodic Table?

    (A) As atomic number increases, the atomic radius decreases.
    (B) As atomic number increases, the atomic radius increases.
    (C) As atomic number increases, the atomic radius remains the same.
    (D) As atomic number remains the same, the atomic radius increases.
    (E) As atomic number remains the same, the atomic radius decreases.

22. Which kind of atom listed below contains the greatest number of unpaired electrons?

    (A) K
    (B) Sc
    (C) V
    (D) Mn
    (E) Co

23. Which describes the gas produced when dilute hydrochloric acid is added to solid sodium hydrogen sulfite, $NaHSO_3$?

    (A) dark brown in color
    (B) turns limewater cloudy
    (C) re-ignites a glowing splint
    (D) has a sharp, choking odor
    (E) has the odor of rotten eggs

24. Which pair gives correct formulas for two different chlorides of iron?

    (A) $FeCl$, $FeCl_2$

    (B) Fe_2Cl_2, $FeCl_2$

    (C) $FeCl_2$, $FeCl_3$

    (D) $FeCl_2$, $FeCl_4$

    (E) $FeCl_3$, $FeCl_4$

25. Which is most soluble in water?

    (A) C_2H_5OH, ethanol

    (B) C_2H_5Cl, chloroethane

    (C) $C_2H_5C_6H_5$, ethyl benzene

    (D) $C_2H_5OC_2H_5$, diethyl ether

    (E) $C_2H_5COOC_2H_5$, ethyl propanoate

Free-Response Questions

26. The following observations, A-D, are made about reactions of sodium hydroxide, NaOH. Discuss the chemical processes involved in each case. Use principles from acid-base theory, oxidation-reduction and bonding and/or intermolecular forces to support your answers.

(A) When a few drops of 3 M NaOH solution are added to 10 mL of 0.50 M aluminum chloride, $AlCl_3$, a white precipitate forms. When excess NaOH solution is added to the mixture containing the precipitate, the precipitate dissolves.

(B) When pellets of NaOH are added to water, there is substantial increase in the temperature of the system as the pellets dissolve.

(C) When 15 mL of 1 M sodium hydroxide, NaOH, is added to 10 mL of 1 M hydrochloric acid, HCl, the resulting mixture has pH greater than 7. When 15 mL of 1 M NaOH is added to 10 mL of 1 M phosphoric acid, H_3PO_4, the resulting mixture has pH less than 7.

(D) When a solution of 0.10 M NaOH is exposed to the atmosphere for several days, both its volume and its pH decrease.

27. A student is given samples of five separate solutions, each containing equal molarities of one of these salts:

$AgNO_3$
NaCl
Na_2S
$KMnO_4$
$CuSO_4$

The samples are unmarked. The student is assigned to identify each solution. No external devices or additional materials are allowed. Using observations of the individual solutions and by mixing them in pairs, describe a strategy that permits identification of each solution. Tasting is not permitted.

28. Explain each observation in terms of chemical reactions.

(A) Solid $Zn(OH)_2$ will dissolve in concentrated HCl or in concentrated NaOH.

(B) Bubbles and heat are generated when sodium metal is added to water at room temperature but neither is observed when aluminum is added to room temperature water.

(C) A dilute solution of $Cu(NO_3)_2$ is a light blue color but addition of a solution of ammonia causes a dark blue solution.

(D) A sample of solid $CuSO_4 \cdot 5H_2O$ is blue initially but turns white after being heated in a crucible for five minutes over a Bunsen burner.

SAMPLE EXAMINATION I

Section I – Multiple Choice

Questions 1–5: The set of lettered choices is a list of the oxides of five different elements. The numbered statements immediately following refer to that list. Select the one lettered choice from the list that best fits each statement. A choice may be used once, more than once, or not at all.

(A) Na_2O

(B) MgO

(C) Al_2O_3

(D) SO_2

(E) Cl_2O

Identify the oxide

1. of the element that exhibits the highest oxidation number

2. that is most soluble in water, forming a strong base

3. that is most closely associated with acid rain

4. that dissolves in water to form a weak molecular monoprotic acid

5. whose Lewis structure indicates the existence of resonance structures

---

Questions 6–10: Each question below refers to a mixture prepared by adding 100 mL of 0.10 M $AgNO_3$ to a 500 mL beaker containing 100 mL of 0.10 M Na_2CrO_4. A precipitate forms in this mixture.

6. Which describes the contents of the beaker before and after the mixture is prepared?

	before	after
(A)	colorless solution	red precipitate
(B)	yellow solution	red precipitate
(C)	yellow solution	yellow precipitate
(D)	colorless solution	yellow precipitate
(E)	colorless solution	white precipitate

7. What is the formula of the precipitate?

   (A) Ag_2O

   (B) Cr_2O_3

   (C) $NaNO_3$

   (D) Ag_2CrO_4

   (E) $NaNO_3$

8. What is $[Na^+]$ in the liquid phase of the mixture?

   (A) 0.040 M

   (B) 0.050 M

   (C) 0.060 M

   (D) 0.10 M

   (E) 0.20 M

9. What is the number of moles of Ag^+ in the precipitate?

   (A) 0.010

   (B) 0.0050

   (C) 0.0025

   (D) 0.0010

   (E) 0.00050

10. Which series lists the three ions, chromate, sodium and nitrate, in order of increasing concentration in the solution after the precipitation reaction has occurred?

    (A) CrO_4^{2-} Na^+ NO_3^-

    (B) CrO_4^{2-} NO_3^- Na^+

    (C) NO_3^- Na^+ CrO_4^{2-}

    (D) Na^+ CrO_4^{2-} NO_3^-

    (E) Na^+ NO_3^- CrO_4^{2-}

---

11. Adding some substances to water causes large amounts of energy to be liberated. All of the following substances illustrate that behavior EXCEPT

    (A) pellets of KOH

    (B) concentrated H_2SO_4

    (C) metallic Na

    (D) crystals of NH_4NO_3

    (E) anhydrous $CuSO_4$

12. Which is the correct comparison of the Cl^0 atom to the Cl^- ion?

    I.   The radius of the Cl^0 atom is greater than the radius of the Cl^- ion.

    II. The mass of the Cl^0 atom is about 1 amu greater than the mass of the Cl^- ion.

    III. The Cl^0 atom contains fewer electrons than the Cl^- ion.

(A) I only

(B) III only

(C) I and II only

(D) II and III only

(E) I, II, and III

13. What is the number of shared pairs of electrons in a molecule of propyne, C_3H_4?

(A) three

(B) four

(C) six

(D) seven

(E) eight

14. In which series are the elements listed in order of increasing atomic radius?

(A) $_{35}Br$, $_{17}Cl$, $_9F$

(B) $_{11}Na$, $_{19}K$, $_{37}Rb$

(C) $_{36}Kr$, $_{18}Ar$, $_{10}Ne$

(D) $_{11}Na$, $_{12}Mg$, $_{13}Al$

(E) $_{34}Se$, $_{16}S$, $_8O$

15. What is the number of occupied orbitals in the third principal energy level of a manganese atom in the ground state?

(A) three

(B) four

(C) five

(D) eight

(E) nine

16. The electron configuration of an atom of element X is $1s^2 2s^2 2p^5$. Which is the best electron dot diagram for an atom of this element?

    (A) 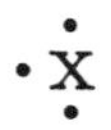 $\cdot \overset{\cdot}{\underset{\cdot}{X}}$

    (B)  $\cdot \overset{\cdot \cdot}{X}$

    (C)  $\cdot \overset{\cdot}{\underset{\cdot}{X}} \cdot$

    (D) 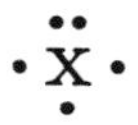 $\cdot \overset{\cdot \cdot}{\underset{\cdot}{X}} \cdot$

    (E) 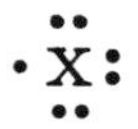 $\cdot \overset{\cdot \cdot}{\underset{\cdot \cdot}{X}} \colon$

17. The term "weighted average atomic mass" refers to a calculated atomic mass that takes into account

    (A) ionization energy
    (B) number of positive valences
    (C) charge on the monatomic ion
    (D) mass defect in the nucleus
    (E) naturally-occurring distribution of isotopes

18. Which set of quantum numbers matches the spectographic notation, $3d^4$?

    (A) $3, 3, 4, -\frac{1}{2}$
    (B) $3, 2, 3, -\frac{1}{2}$
    (C) $3, 2, 1, +\frac{1}{2}$
    (D) $4, 3, 0, -\frac{1}{2}$
    (E) $4, 3, 1, +\frac{1}{2}$

19. What is the number of possible isomers of the complex ion, $[Co(H_2O)_4(NH_3)_2]^{2+}$?

    (A) two
    (B) three
    (C) four
    (D) six
    (E) eight

20. Which classifies all the bonds in a molecule of methyl methanoate, $HCOOCH_3$?

    (A) 3 *sigma* bonds and 1 *pi* bond

    (B) 4 *sigma* bonds and 2 *pi* bonds

    (C) 5 *sigma* bonds and 2 *pi* bonds

    (D) 6 *sigma* bonds and 1 *pi* bond

    (E) 7 *sigma* bonds and 1 *pi* bond

21. Which of the following is closest to the measurement of the $O-C-O$ bond angle in methyl methanoate, $HCOOCH_3$?

    (A) 60°

    (B) 90°

    (C) 109°

    (D) 120°

    (E) 180°

22. Which is the most likely formula for hydrogen arsenide?

    (A) HAs

    (B) HAs_2

    (C) H_2As

    (D) H_3As_2

    (E) H_3As

23. The symbols, $_1^1H$, $_1^2H$ and $_1^3H$, represent three different

    (A) homologs

    (B) isotopes

    (C) isomers

    (D) allotropes

    (E) conformations

24. Which element has the greatest difference between its first and second ionization energies?

    (A) bromine

    (B) calcium

    (C) germanium

    (D) potassium

    (E) scandium

25. In the methyl ethanoate molecule, CH_3COOCH_3, for which other bond angle is its measurement closest to the measurement of its $O-C-O$ bond angle?

    (A) $H-C-O$
    (B) $C-C-O$
    (C) $H-C-H$
    (D) $H-C-C$
    (E) $C-O-C$

26. All of the following apply to bonding in the PF_5 molecule EXCEPT

    (A) d^2sp^3 hybridization
    (B) $F-P-F$ bond angles of 90°
    (C) $F-P-F$ bond angles of 120°
    (D) expanded octet of electrons
    (E) trigonal bipyramidal geometry

27. Which pair of formulas is most closely associated with the Law of Multiple Proportions?

    (A) $CuCl$ and $CuCl_2$

    (B) C_2H_2 and C_2H_6

    (C) $CuBr_2$ and CuI_2

    (D) $C_2H_5OC_2H_5$ and C_4H_9OH

    (E) CH_2Br_2 and $CHBr_3$

28. Which term completes this unbalanced nuclear equation?

    $$^{238}_{94}Pu + 2\,^{1}_{0}n \rightarrow \ \_?\_ \ + \_^{0}_{-1}e$$

    (A) $^{237}_{93}Np$

    (B) $^{239}_{93}Np$

    (C) $^{239}_{95}Am$

    (D) $^{240}_{93}Np$

    (E) $^{240}_{95}Am$

29. What is the best description of the species found at the lattice points of the mineral calcite, $CaCO_3$?

    (A) $CaCO_3$ molecules

    (B) Ca^{2+} ions and $CO_3{}^{2-}$ ions

    (C) Ca atoms, C atoms and O atoms

    (D) Ca^{2+} ions, C^{4+} ions and O^{2-} ions

    (E) CaO molecules and CO_2 molecules

30.

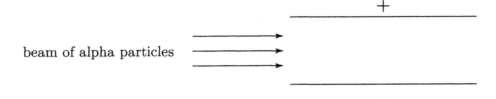

beam of alpha particles

Which describes the behavior of a beam of alpha particles as it passes through the electric field between two charged plates as shown above?

    (A) The beam passes through unchanged.

    (B) The beam is destroyed by the electric field.

    (C) The beam is deflected back toward its origin.

    (D) The beam is deflected toward the negative plate.

    (E) The beam is deflected toward the positive plate.

31. Which property is the same for any two samples of $SO_{2(g)}$ and $SO_{3(g)}$ at the same temperature?

    (A) critical temperature

    (B) number of molecules

    (C) average kinetic energy

    (D) pressure-volume product

    (E) average molecular velocity

32. The molar mass of an unidentified gas is 68 g. Assuming ideal behavior, its density in grams per liter at STP is closest to

    (A) 1.0

    (B) 1.5

    (C) 2.0

    (D) 2.5

    (E) 3.0

33. Consider a sample of gas confined at constant temperature and pressure in the piston system shown below. If more of this same gas is added to the piston at constant temperature, what effect is observed on volume and average molecular velocity?

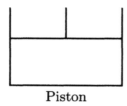

Piston

(A) Both volume and average molecular velocity increase.

(B) Both volume and average molecular velocity remain the same.

(C) Volume remains the same and average molecular velocity increases.

(D) Volume remains the same and average molecular velocity decreases.

(E) Volume increases and average molecular velocity remains the same.

34. Consider the three sealed identical flasks represented below, each containing 0.100 mole of the gas specified at 1 atm and 273 K.

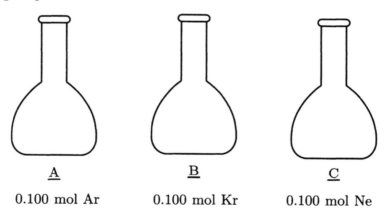

A   B   C

0.100 mol Ar   0.100 mol Kr   0.100 mol Ne

Which is a correct comparison of the contents of the flasks?

   I.   The mass of the contents of flask B is the greatest.

   II. The number of molecules in each flask is the same.

   III. The density of the contents of each flask is the same.

(A) I only

(B) II only

(C) I and II only

(D) II and III only

(E) I, II, and III

35. A system is prepared by sealing a flask half-filled with 1.0 m solution of NaCl. The system is kept at constant temperature. Which description applies to the processes of condensation and evaporation of water in that system?

    (A) Both evaporation and condensation occur continuously.

    (B) Since the flask is sealed, neither evaporation nor condensation occurs.

    (C) Only evaporation occurs, which stops when equilibrium is achieved.

    (D) Only condensation occurs, which stops when equilibrium is achieved.

    (E) Both evaporation and condensation occur, stopping when the air above the solution is saturated with $H_2O_{(g)}$.

36. The van der Waals equation of state for real gases is

$$\left(P + \frac{n^2 a}{V^2}\right)(V - nb) = nRT$$

    where the values for the <u>a</u> and <u>b</u> coefficients have been determined experimentally. Which property of the molecules in a sample is most closely related to the value of the <u>a</u> coefficient?

    (A) the mass of the molecules

    (B) the volume of the molecules

    (C) number of molecules in the sample

    (D) forces of attraction between molecules

    (E) the root mean velocity of the molecules

37. Which applies to a dilute solution of sodium chloride in water?

    I.   Adding sodium chloride lowers the freezing point.

    II. Adding sodium chloride decreases the vapor pressure of the solution.

    III. Adding sodium chloride decreases the density of the solution.

    (A) II only

    (B) III only

    (C) I and II only

    (D) I and III only

    (E) I, II, and III

Questions 38 and 39: The phase diagram below represents a hypothetical substance.

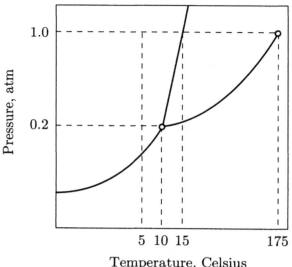

38. Which temperature range includes the boiling point of the substance at an elevation 5,000 feet above sea level?

(A) greater than 175°C

(B) 100° to 175°C

(C) 15°C to 100°C

(D) −5° to 15°C

(E) less than −15°C

39. Which gives correct information about the substance as represented in the phase diagram?

I. The vapor pressure of the liquid phase at 15°C is less than 0.5 atm.

II. At any pressure less than 0.2 atm, the solid undergoes sublimation.

III. The density of the solid phase is greater than the density of the liquid phase.

(A) II only

(B) III only

(C) I and III only

(D) II and III only

(E) I, II, and III

40. All of the following are colligative properties of a solution EXCEPT

   (A) boiling point elevation

   (B) freezing point depression

   (C) molar conductance

   (D) osmotic pressure

   (E) vapor pressure

41. Consider the list of nitrogen-containing species below.

   $$N^{3-}, NH^{2-}, NH_2^{-}, NO_3^{-}, NO_2^{-}, NH_3, NH_4^{+}$$

   Which answer includes all of the species from the list above that cannot behave as Lewis bases?

   (A) N^{3-} only

   (B) NH_4^{+} only

   (C) NH_3 only

   (D) NO_3^{-} and NO_2^{-} only

   (E) N^{3-}, NH^{2-} and NH_2^{-} only

42. Which observations support the claim that $Al(OH)_3$ behaves as an amphoteric substance?

   I.   The solid present in a suspension of $Al(OH)_3$ dissolves upon the addition of $HCl_{(aq)}$.

   II. The solid present in a suspension of $Al(OH)_3$ dissolves upon the addition of $NaOH_{(aq)}$.

   III. No change is observed when NaCl is added to a suspension of $Al(OH)_3$.

   (A) I only

   (B) I and II only

   (C) I and III only

   (D) II and III only

   (E) I, II, and III

43. A quantity of liquid solution, specified as 20 mL, is to be added to a reaction mixture. Which vessel provides the most precise measurement of the volume of that liquid?

   (A) 25 mL volumetric flask

   (B) 25 mL graduated cylinder

   (C) 25 mL volumetric pipette

   (D) 50 mL buret

   (E) 50 mL Erlenmeyer flask

Questions 44-47: Consider the electrochemical cell represented below using the following reduction half-reactions and their E° values:

$$Fe^{3+} + e^- \rightarrow Fe^{2+} \qquad E° = 0.77 \text{ volts}$$
$$Pb^{2+} + 2e^- \rightarrow Pb^0 \qquad E° = -0.13 \text{ volts}$$

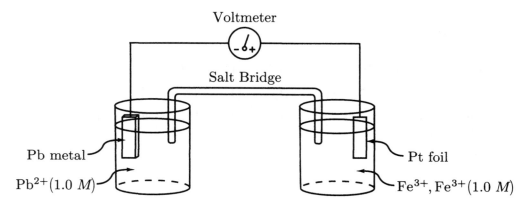

44. Which describes change in concentration of Pb^{2+} and the movement of charge in this electrochemical cell as the cell undergoes discharge?

	movement of electrons in the external circuit	movement of positive ions in the salt bridge	change in $[Pb^{2+}]$
(A)	toward the cathode	toward the cathode	increases
(B)	toward the anode	toward the anode	increases
(C)	toward the cathode	toward the anode	decreases
(D)	toward the anode	toward the cathode	decreases
(E)	toward the cathode	toward the anode	increases

45. Which expression gives the change in mass expected at the lead electrode after this cell has produced 150 milli-amps for 2.0 hours?

(A) $\dfrac{3,600 \times 207}{0.150 \times 96,500}$

(B) $\dfrac{0.150 \times 3,600 \times 207}{2 \times 96,500}$

(C) $\dfrac{2 \times 0.150 \times 207}{96,500}$

(D) $\dfrac{3,600 \times 0.150 \times 207}{96,500}$

(E) $\dfrac{2 \times 0.150 \times 3,600 \times 207}{96,500}$

46. Which expression gives the voltage for this standard chemical cell?

    (A) $0.13 + 0.77$ volts

    (B) $-0.13 + 0.77$ volts

    (C) $0.13 + (2 \times 0.77)$ volts

    (D) $(2 \times 0.13) + (2 \times 0.77)$ volts

    (E) $\left(2 \times (-0.13)\right) + (2 \times 0.77)$ volts

47. A similar electrochemical cell is assembled using standard electrodes except that the concentration of Pb^{2+} is changed to $0.010\ M$. Which is the best comparison of the voltage of the original standard cell to this non-standard cell?

    (A) No difference is expected.

    (B) The voltage increases by about 0.06 volts.

    (C) The voltage decreases by about 0.06 volts.

    (D) The voltage increases by about 0.12 volts.

    (E) The voltage decreases by about 0.12 volts.

---

48. How many moles of methanol should be added to 6.0 moles of water to produce a solution that is 0.75 mole fraction in methanol?

    (A)   2.0

    (B)   8.0

    (C) 12

    (D) 15

    (E) 18

49. Which is the best description of the change that occurs when $Na_2O_{(s)}$ is dissolved in water?

    (A) The oxide ion accepts a share in a pair of electrons.

    (B) The oxide ion donates a share in a pair of electrons.

    (C) The oxidation number of oxygen increases.

    (D) The oxidation number of sodium increases.

    (E) The oxidation number of sodium decreases.

50. Which species is present at the lowest concentration in a $0.10\ M$ solution of $KHSO_4$?

    (A) K^+

    (B) HSO_4^-

    (C) SO_4^{2-}

    (D) H_3O^+

    (E) H_2SO_4

51. Compared to the value of ΔH_f° for $H_2O_{(s)}$, the value of ΔH_f° for $H_2O_{(\ell)}$ has the

    (A) opposite sign and the same absolute value

    (B) same sign and smaller absolute value

    (C) same sign and greater absolute value

    (D) opposite sign and greater absolute value

    (E) opposite sign and smaller absolute value

52. In an ordinary alkaline flashlight cell, zinc metal is oxidized to Zn^{2+} as an electromotive force (E°) of 1.5 volts is produced. The value for $\Delta G°$ in kilojoules per mole of Zn oxidized in this cell is closest to

    (A) 300

    (B) 150

    (C) 0

    (D) −150

    (E) −300

53. Consider two sealed flasks with different volumes each containing 0.10 mol of gas at the same temperature as shown below.

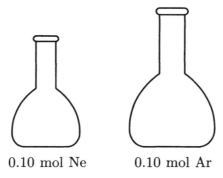

0.10 mol Ne   0.10 mol Ar

Which is a correct comparison of the contents of these flasks?

     I.   The mass of the contents of each flask is the same.

    II.   The number of molecules in each flask is the same.

    III.   The average molecular velocity of the molecules in each flask is the same.

(A) I only

(B) II only

(C) I and II only

(D) II and III only

(E) I, II and III

54. Which of the following oxides of nitrogen has the greatest percent oxygen by mass?

(A) NO

(B) NO_2

(C) N_2O

(D) N_2O_4

(E) N_2O_5

55. $$2Na_2O_2 + 2H_2O \rightarrow 4NaOH + O_2$$

Which expression gives the mass of O_2 produced when 15.0 g Na_2O_2 (molar mass: 78 g) reacts with water, according to the equation above?

(A) $15.0 \times \dfrac{78}{1} \times \dfrac{1}{2} \times \dfrac{32}{1}$

(B) $15.0 \times \dfrac{1}{78} \times \dfrac{1}{2} \times \dfrac{32}{1}$

(C) $15.0 \times \dfrac{1}{78} \times \dfrac{1}{2} \times \dfrac{1}{32}$

(D) $15.0 \times \dfrac{1}{78} \times \dfrac{2}{1} \times \dfrac{1}{32}$

(E) $15.0 \times \dfrac{78}{1} \times \dfrac{1}{2} \times \dfrac{32}{1}$

Questions 56 and 57: Nitrous acid ionizes according to the equation below:

$$HNO_{2(aq)} \rightleftharpoons H^+{}_{(aq)} + NO_2{}^-{}_{(aq)} \qquad K_a = 7.2 \times 10^{-4}$$

56. Which occurs as more water is added to an equilibrium system of nitrous acid at constant temperature?

total number of cations and anions	percent ionization of HNO_2 molecules
(A) decreases	increases
(B) decreases	decreases
(C) decreases	remains the same
(D) increases	increases
(E) increases	remains the same

57. Which occurs when equal volumes of 1.0 M $NaNO_2$ and 1.0 M HNO_2 are mixed together in a suitable container?

[HNO_2]	total number of ions in solution	percent ionization of HNO_2 molecules
(A) decreases	increases	decreases
(B) remains the same	increases	remains the same
(C) decreases	increases	remains the same
(D) remains the same	decreases	decreases
(E) remains the same	decreases	increases

---

58. Given $K_a = 1.8 \times 10^{-5}$ and $[H^+] = 4.2 \times 10^{-3}$ for 1.0 M CH_3COOH at 298 K. As temperature increases, the percent ionization increases. Which describes the corresponding effects on pK_a and pH?

pK_a	pH
(A) remains the same	increases
(B) remains the same	decreases
(C) increases	increases
(D) decreases	decreases
(E) decreases	increases

59. The K_{sp} for AgI at 25°C is 8.3×10^{-17}. Which expression is closest to the molar solubility of AgI?

    (A) 9×10^{-8}

    (B) 9×10^{-9}

    (C) $\frac{8.3}{235} \times 10^{-9}$

    (D) $\frac{235}{8.3} \times 10^{-8}$

    (E) $8.3 \times 235 \times 10^{-8}$

60. $$BaSO_4 \rightleftharpoons Ba^{2+} + SO_4^{2-} \qquad\qquad K_{sp} = 1 \times 10^{-10}$$

    Which range includes the minimum number of moles of $BaCl_2$ that must be added to 1.0 liter of a saturated solution of $BaSO_4$ to change $[SO_4^{2-}]$ to 5×10^{-10} M?

    (A) less than 0.1×10^{-10}

    (B) from 1×10^{-10} to 6×10^{-10}

    (C) from 6×10^{-10} to 6×10^{-6}

    (D) from 6×10^{-6} to 6×10^{-3}

    (E) more than 6×10^{-3}

61. Which of the following has the least effect on the rate of a reaction in the gas phase?

    (A) adding a solid surface catalyst
    (B) adding inert gas at constant volume
    (C) decreasing the volume of the reaction system
    (D) increasing the temperature of the reaction system
    (E) decreasing the concentration of one of the reactants

62. In any first order reaction, as the reaction proceeds at constant temperature, which describes the corresponding effects on k (the rate constant) and *rate*?

	k	*rate*
(A)	remains the same	decreases
(B)	remains the same	remains the same
(C)	remains the same	increases
(D)	decreases	decreases
(E)	decreases	remains the same

Questions 63 and 64: The steps below represent a mechanism proposed for the reaction of nitrogen(II) oxide with hydrogen.

$$NO_{(g)} + NO_{(g)} \rightleftharpoons N_2O_{2(g)} \qquad \text{fast equilibrium}$$

$$N_2O_{2(g)} + H_{2(g)} \rightleftharpoons N_2O_{(g)} + H_2O_{(g)} \qquad \text{slow}$$

$$N_2O_{(g)} + H_{2(g)} \rightleftharpoons N_{2(g)} + H_2O_{(g)} \qquad \text{fast}$$

63. Which rate law is consistent with this mechanism?

(A) rate $= k[NO]^2$

(B) rate $= k[NO]^2[H_2]$

(C) rate $= k[N_2O][H_2]$

(D) rate $= k[N_2O_2][H_2]$

(E) rate $= k[N_2O_2][H_2][NO]^2$

64. Which is the equation for the overall reaction?

(A) $NO_{(g)} + NO_{(g)} \rightarrow N_{2(g)} + H_2O_{(g)}$

(B) $2NO_{(g)} + 2H_{2(g)} \rightarrow N_{2(g)} + 2H_2O_{(g)}$

(C) $NO_{(g)} + N_2O_{(g)} + H_{2(g)} \rightarrow N_{2(g)} + H_2O_{(g)}$

(D) $2NO_{(g)} + 2H_{2(g)} + N_2O_{2(g)} \rightarrow N_2O_{(g)} + N_{2(g)} + 2H_2O_{(g)}$

(E) $NO_{(g)} + 2H_{2(g)} + N_2O_{2(g)} + N_2O_{(g)} \rightarrow N_2O_{(g)} + N_2O_{2(g)} + N_{2(g)} + 2H_2O_{(g)}$

---

65. Which characteristic of a closed reaction system is most affected by the presence of a catalyst?

(A) free energy change

(B) enthalpy of reaction

(C) standard entropy of formation

(D) time required to reach equilibrium

(E) percent conversion to products at equilibrium

66. Which describes the direction of changes in enthalpy and entropy for an endothermic dissolving process of an ionic solute in water at constant temperature?

(A) Enthalpy increases and entropy decreases.

(B) Enthalpy decreases and entropy increases.

(C) Both enthalpy and entropy decrease.

(D) Both enthalpy and entropy increase.

(E) Enthalpy decreases and entropy remains the same.

67. Which of the following is most closely associated with relatively slow rates of chemical reaction?

(A) low enthalpy of reaction

(B) low energy of activation

(C) the presence of a catalyst

(D) high concentration of reactants

(E) strong bonds in reaction molecules

68. Which applies to any changes in entropy and enthalpy associated with the vaporization of any liquid at any temperature?

(A) Both entropy and enthalpy increase.

(B) Entropy increases and enthalpy decreases.

(C) Entropy remains the same and enthalpy increases.

(D) Entropy remains the same and enthalpy decreases.

(E) Entropy increases and enthalpy remains the same.

69. Consider the reaction system for the endothermic decomposition below at equilibrium in a 2.0 liter sealed rigid flask at 298 K.

$$PH_3PCl_{3(s)} \rightleftharpoons PH_{3(g)} + PCl_{3(g)} \qquad K_{eq} = 3.5 \times 10^{-2}$$

When the vessel containing the system is immersed in an ice bath, all of the following occur EXCEPT

(A) The total pressure decreases.

(B) The volume of $PH_{3(g)}$ decreases.

(C) The total number of all molecules decreases.

(D) The partial pressure of $PCl_{3(g)}$ decreases.

(E) The number of molecules of $PH_3PCl_{3(s)}$ increases.

70. Which process is accompanied by a decrease in entropy?

(A) melting of a metallic solid

(B) dissolving of an ionic solid

(C) evaporation of a molecular liquid

(D) increase in volume of a confined sample of gas

(E) formation of a crystalline solid from a supersaturated solution

71. In an experiment to determine the concentration of a solution of hydrochloric acid, a worker placed precisely 25.0 mL of the unknown acid solution in a beaker that contained about 50 mL of water and 4 drops of phenolphthalein solution. The worker then filled a 50 mL buret to the zero mark with 0.300 M NaOH solution and began titration immediately. Which describes a mistake in this procedure?

    I. Because phenolphthalein does not change color until $[H^+]$ is about 10^{-9}, this substance is not a suitable indicator for this experiment.

    II. Because the worker added the unknown acid solution to an unmeasured quantity of water, the concentration of the unknown acid solution cannot be determined.

    III. Because the worker did not fill the tip of the buret before beginning the titration, an accurate measurement of the volume of base used cannot be obtained.

    (A) I only

    (B) II only

    (C) III only

    (D) I and III only

    (E) I, II, and III

72. Which compound of potassium is a colored solid?

    (A) $KBrO_3$

    (B) $KAl(SO_4)_2$

    (C) K_2PtCl_6

    (D) K_2SeO_3

    (E) $K_2Si_2O_5$

73. Which is the best description of the concentration of ions in solution when 0.050 mol $OH^-_{(aq)}$ is added to 1.0 liter of 0.10 M solution of NaH_2PO_4? (Assume no change in volume.)

	$[H_2PO_4{}^-]$	$[HPO_4{}^{2-}]$	$[PO_4{}^{3-}]$
(A)	negligibly small	0.050	negligibly small
(B)	negligibly small	negligibly small	0.050
(C)	0.050	0.050	negligibly small
(D)	0.050	negligibly small	0.050
(E)	0.075	negligibly small	0.025

74. The essential elements in every amino acid include each of the following EXCEPT

    (A) carbon

    (B) oxygen

    (C) nitrogen

    (D) hydrogen

    (E) phosphorus

75. When 10 mL of 3.0 M $NH_{3(aq)}$ is added to 100 mL of a colorless solution, the change in pH is observed to be less than 0.10 of a pH unit. The colorless solution could have been.

    I.   3 M HCl
    II. 3 M NaOH
    III. 3 M NaCl

    (A) I only

    (B) II only

    (C) I and II only

    (D) I and III only

    (E) I, II, and III

Section II

Section II - Free Response   Total Time – 90 Minutes
(Multiple-Choice Questions are found in Section I.)

Part A: Question 76
and
Question 77 or Question 78
Time: 40 minutes

Access to calculators, Periodic Table, lists of standard reduction potentials, and
Equations and Constants

(2004 Examination directions) Clearly show the method used and the steps involved in arriving at your answers. It is to your advantage to do this, because you may obtain partial credit if you do and you will receive little or no credit if you do not. Attention should be paid to significant figures. Be sure to write all your answers to the questions on the lined pages following each question in the booklet with the pink cover. Do not write your answers on the green insert.

Answer question 76 below. The Section II score weighting for this question is 20 percent.

76.   Methylamine, CH_3NH_2, is an organic base, accepting a proton from water to form the methylammonium ion. The value of K_{eq} for this system is 4.0×10^{-4} at 298 K.

(a)   Write the chemical equation for the equilibrium as described above.

(b)   Calculate the concentration of hydroxide ions in a 0.25 M solution of methylamine.

(c)   How is the equilibrium affected when solid NaOH is added to a solution of methylamine? Calculate the $[CH_3NH_3^+]$ when 0.020 mol OH^- is added to 500. mL of 0.25 M methylamine, CH_3NH_2. (Assume no change in volume.)

(d)   A buffer solution is prepared that is 0.20 M in $CH_3NH_3^+$ and 0.25 M in CH_3NH_2. Calculate the pH of this solution.

(e)   Calculate the number of moles of H^+ that must be added to 200. mL of the solution in part (d) in order to change the pH to 10.00.

Answer either question 77 or question 78 below.

(2004 examination directions) Only one of these two questions will be graded. If you start both questions, be sure to cross out the question you do not want graded.

The Section II score weighting for the question that you choose is 20 percent.

77.   Answer the following questions about a chromium/hydrogen electrochemical cell.

(a)   Make a labeled sketch of an electrochemical cell using a standard Cr/Cr^{3+} half cell connected to a standard hydrogen half-cell. Your labels should include
- anode
- cathode
- chemical components and concentration(s) in the chromium half cell
- direction of electron flow in the external circuit
- path for ion migration.

Include a labeled voltmeter in the external circuit.

(b)   Write the half reactions and the balanced overall equation for this cell.

(c)   Calculate the voltage for this standard cell.

(d)   Calculate the voltage when the concentration of Cr^{3+} is 0.050 M.

78.   Answer all four questions below about the samples of gases described. Consider separate samples of H_2S gas and SO_2 gas. The mass of each sample is 10.0 g.

(a)   What is the ratio of the number of sulfur atoms in the sample of $H_2S_{(g)}$ compared to the number of sulfur atoms in the sample of $SO_{2(g)}$?

(b)   What is the volume occupied by the H_2S gas, when measured at 25°C and 745 mm Hg?

(c)   Calculate the ratio of the average velocity of H_2S gas molecules to that of SO_2 gas molecules when both samples are measured at the same temperature.

(d)   Separate sources of $H_2S_{(g)}$ and $SO_{2(g)}$ are placed in opposite ends of a 100 cm tube. When these gases meet, they react to form solid sulfur. At what distance from the SO_2 end will the gases meet and the deposit of solid sulfur first be observed?

Part B: Questions 79, 80, 81 and
Question 82 or Question 83
Time: 50 minutes

Access to Periodic Table, lists of standard reduction potentials
and *Equations and Constants*
No access to calculators

Answer question 79 below: The Section II score weighting for this question is 15 percent.

79. (2004 Examination directions) Write the formulas to show the reactants and products for FIVE of the laboratory situations described below. Answers to more than five choices will not be graded. In all cases a reaction occurs. Assume that solutions are aqueous unless otherwise indicated. Represent substances in a solution as ions if the substances are extensively ionized. Omit formulas for any ions or molecules that are unchanged by the reaction. You need not balance the equations.

(a) Excess hydrochloric acid is added to a solution of sodium hydrogen phosphate.

(b) A sample of solid lithium oxide is added to water.

(c) Hydrogen peroxide solution is added to a solution of iron(II) chloride.

(d) A solution of potassium iodide is electrolyzed.

(e) A strip of copper is immersed in dilute nitric acid.

(f) A few crystals of calcium fluoride are added to hot concentrated sulfuric acid.

(g) A sample of ethanol is ignited in excess oxygen.

(h) Solutions of ammonium thiocyanate and iron(III) chloride are mixed.

(2004 Examination directions) Your responses to the rest of the questions in this part of the examination will be graded on the basis of the accuracy and relevance of the information cited. Explanations should be clear and well organized. Examples and equations may be included in your responses where appropriate. Specific answers are preferable to broad, diffuse responses.

(2004 examination directions) **Answer both Question 80 and Question 81 below.** Both questions will be graded.

The Section II score weighting for these questions is 30 percent (15 percent each).

80. Periodic Relationships

(a) Ionization Energies

Ionization energies, kJ mol^{-1}

	Na	Mg	Al
First Ionization Energy	496	738	578
Second Ionization Energy	4,560	1,450	1,820
Third Ionization Energy	6,917	7,730	2,750

(i) The second ionization for each element is greater than the first ionization energy for that element. Explain.

(ii) The difference between first and second ionization energies is much greater for Na than for Mg. Explain.

(b) Atomic/Ionic Radius

	$_{16}S$	$_{16}S^{2-}$	$_{20}Ca$	$_{20}Ca^{2+}$
Atomic/ionic radius, nm	0.104	0.184	.197	0.099

(i) The radius of $_{16}S$ is less than the radius of $_{16}S^{2-}$. Explain.

(ii) The $_{16}S^{2-}$ and $_{20}Ca^{2+}$ are isoelectronic species. However, the radius of $_{16}S^{2-}$ is greater than the radius of $_{20}Ca^{2+}$. Explain.

81. Answer all four questions about the laboratory procedures below.
The questions below are related to the exothermic dissolving of $CaCl_{2(s)}$ in water.

(a) Describe the energy changes that occur as $CaCl_{2(s)}$ dissolves.

(b) Describe how to use 0.50 mol (55.5 g) of $CaCl_{2(s)}$ to make each of the aqueous solutions specified below. For each, specify the mass or volume of the solution produced and specify how the amount of liquid solvent or solution is to be measured.

(i) a quantity of 1.0 molar (1.0 M) solution
(ii) a quantity of 1.0 molal (1.0 m) solution

(c) Describe how to use solid $CaCl_2$ to determine its heat of solution in kJ mol^{-1}. A styrofoam cup is available to use as a calorimeter. A thermometer, a balance, water and ordinary lab equipment are also available. Specify the procedure and the measurements to be recorded. (Calculation or description of calculation is not required.)

(2004 examination directions) **Answer either question 82 or question 83 below.** Only one of these two questions will be graded. If you start both questions, be sure to cross out the question you do not want graded. The Section II score weighting for the question that you choose is 15 percent.

82. Answer all four questions about the burning of octane.

$$C_8H_{18(\ell)} + \frac{25}{2}O_{2(g)} \rightarrow 8CO_{2(g)} + 9H_2O_{(g)} + heat$$

The combustion reaction above is the source of the energy produced by the burning of octane in an automobile engine. This reaction is spontaneous at 298 K.

(a) Predict the sign of ΔS in the reaction. Explain.

(b) Predict the sign for ΔG for this reaction at 298 K. Explain.

(c) $\Delta H^\circ_{f,CO_{2(g)}} = -393.5 \text{ kJ mol}^{-1}$; $\Delta H^\circ_{f,CO_{(g)}} = -110.5 \text{ kJ mol}^{-1}$

   If some of the reactants were converted to CO rather than CO_2, how would the total amount of energy produced be affected? Explain.

(d) If this reaction were carried out at a temperature greater than 298 K, for which of the three parameters, ΔH, ΔG or ΔS, would the change in value have the greatest magnitude? Explain.

83. Answer all three questions below about this reaction:

$$A_{(g)} + B_{(g)} \rightarrow AB_{(g)} + energy$$

The rate of the reaction above is known to be first order in A and first order in B.

The rate increases when a suitable catalyst is added.

Use the axis below for the answers to part (a) and (b).

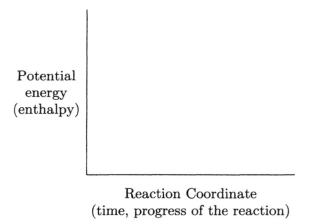

Potential energy (enthalpy)

Reaction Coordinate
(time, progress of the reaction)

(a) On the labeled axes above, draw a diagram of potential energy versus reaction coordinate for the uncatalyzed reaction. Use a line labeled (A) to show the progress of this reaction. On the diagram, label:

(i) relative potential energy of the

   • reactants

   • activated complex

   • product

(ii) the intervals that correspond to

   • heat of reaction, ΔH

   • activation energy for the forward reaction, $E_{a,f}$

   • activation energy for the reverse reaction, $E_{a,r}$

(b) On the set of axes, use a line labeled (B) to show the change or changes in the values in part (a) (i), above, that result from the addition of the catalyst.

(c) At ordinary conditions, $RATE_f$ is much greater than $RATE_r$. The symbols k_f and k_r represent the rate constants for the forward and reverse reactions respectively. How does the value of the ratio $\dfrac{k_f}{k_r}$ change, when temperature increases? Explain.

No testing material printed on this page.

SAMPLE EXAMINATION II

Section I – Multiple Choice

Questions 1-5: The set of lettered choices below is a list of chemical elements. The list refers to the numbered statements immediately following it. Select the one lettered choice that best fits each numbered statement. A choice may be used once, more than once or not at all.

(A) silicon
(B) phosphorus
(C) nitrogen
(D) magnesium
(E) bromine

1. the element whose oxide is a molecular solid at ordinary conditions

2. the element with the lowest melting point

3. the element that is best conductor of electricity

4. the element that is a molecular solid at ordinary conditions

5. the element whose oxide is an ionic solid

Questions 6-10: The set of lettered choices below is a list of formulas for chemical compounds. The list refers to the numbered questions immediately following it. Select the one lettered choice that best answers each question. A choice may be used once, more than once or not at all.

(A) $(NH_4)_2SO_4$
(B) K_2S
(C) CO_2
(D) $CaCl_2$
(E) $NaClO_4$

Which formula represents or includes

6. a cation that precipitates from aqueous solution when a solution containing carbonate anion is added

7. an atom with an oxidation number of -3

8. a molecule that contains two *sigma* (σ) and two *pi* (π) bonds

9. a species which, when heated along with a catalyst, yields a salt containing a halide in addition to oxygen gas.

10. a gas, when measured under standard conditions

Questions 11-16: The set of lettered choices below is a list of categories of hydrocarbon derivatives. It refers to the set of numbered structural formulas immediately following. For each structural formula select the one lettered choice that identifies the category of hydrocarbon derivative illustrated. A choice may be used once, more than once, or not at all.

(A) secondary alcohol
(B) tertiary alcohol
(C) organic acid
(D) ketone
(E) ester

11. 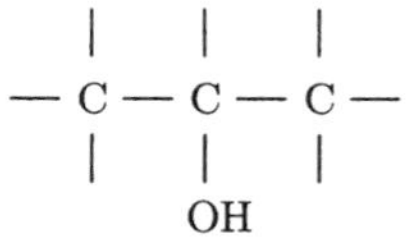
$$-\overset{|}{C}-\overset{|}{\underset{\underset{OH}{|}}{C}}-\overset{|}{C}-$$

14. 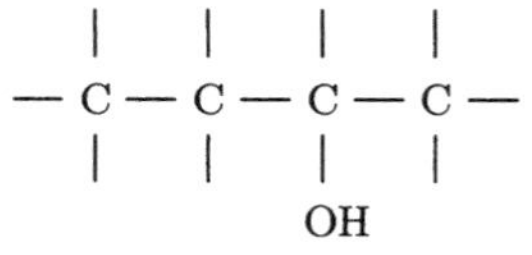
$$-\overset{|}{C}-\overset{|}{C}-\overset{|}{\underset{\underset{OH}{|}}{C}}-\overset{|}{C}-$$

12. 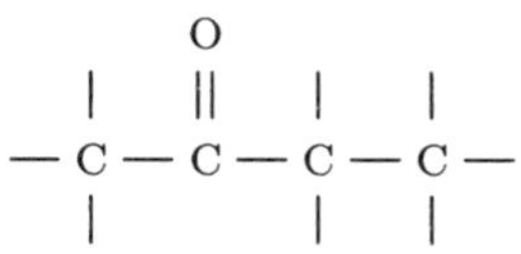
$$-\overset{|}{C}-\overset{\overset{O}{\|}}{C}-\overset{|}{C}-\overset{|}{C}-$$

15. 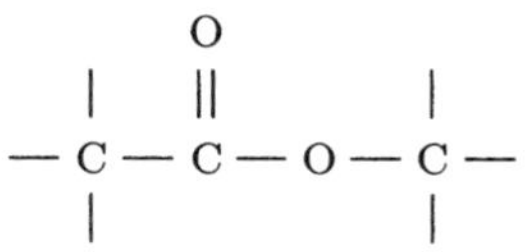
$$-\overset{|}{C}-\overset{\overset{O}{\|}}{C}-O-\overset{|}{C}-$$

13. 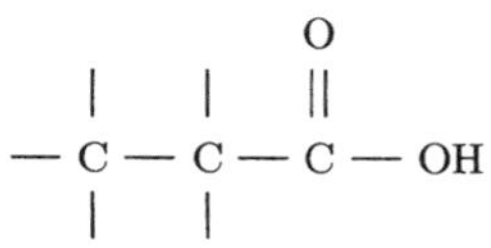
$$-\overset{|}{C}-\overset{|}{C}-\overset{\overset{O}{\|}}{C}-OH$$

16. 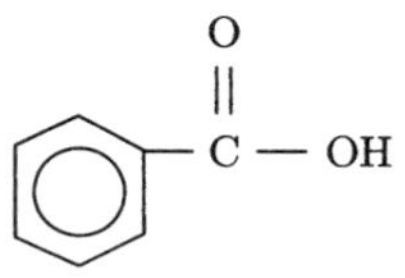
$$\underset{}{\bigcirc}-\overset{\overset{O}{\|}}{C}-OH$$

---

17. All of the following are polar molecules EXCEPT

   (A) H_2S
   (B) NH_3
   (C) PCl_3
   (D) PF_5
   (E) HNO_3

18. Which distribution of electrons in hybrid orbitals is associated with the structure of the sulfur tetrafluoride, SF_4, molecule?

   (A) sp
   (B) sp^2
   (C) sp^3
   (D) dsp^3
   (E) d^2sp^3

19. When each member of the following pairs of solutes is dissolved in separate containers of water, each member of the pair of solutions has a different color EXCEPT for the pair

    (A) $ZnSO_4$ and $CaCl_2$

    (B) $CoCl_2$ and $CrCl_3$

    (C) K_2CrO_4 and $Na_2Cr_2O_7$

    (D) $Ni(NO_3)_2$ and $CuBr_2$

    (E) $AlCl_3$ and $KMnO_4$

20. An atom with the electron configuration $1s^2 2s^2 2p^6 3s^2 3p^6 4s^2 3d^5$ has an occupied but incomplete

    (A) $2s$ sublevel
    (B) $3s$ sublevel
    (C) $3p$ sublevel
    (D) $3d$ sublevel
    (E) $4s$ sublevel

21. The correct name for the compound $(NH_4)_2CuCl_4$ is

    (A) diammine copper(IV) chloride
    (B) diammine tetrachlorocopper
    (C) diammonium copper chlorate
    (D) ammonium tetrachlorocopper
    (E) ammonium tetrachlorocuprate(II)

22. What is the charge on the anion of the compound potassium hexachloroferrate(II)?

    (A) 1^-
    (B) 2^-
    (C) 3^-
    (D) 4^-
    (E) 6^-

23. Which forms cations with charge 1+ and 2+?

    (A) Zn
    (B) Sn
    (C) Fe
    (D) Sc
    (E) Cu

24. In the nitrite anion, NO_2^-, the O-N-O bond angle is slightly less than 120°. Which hybridization of orbitals around the central atom, nitrogen, provides the best explanation for this bond angle.

    (A) sp

    (B) sp^2

    (C) sp^3

    (D) dsp^3

    (E) d^2sp^3

Questions 25 and 26: A mixture of gases is prepared by placing 00.20 mol NO and 0.20 mol NO_2 in a 2.0 liter flask at 300 K. (Assume no chemical reaction occurs.)

25. Which range of values includes the total pressure of the system?

    (A) 1.5 atm to 3.0 atm

    (B) 3.0 atm to 6.0 atm

    (C) 6.0 atm to 12 atm

    (D) 12 atm to 24 atm

    (E) 24 atm to 48 atm

26. Which comparison correctly describes this system?

    I.   The partial pressure of NO is the same as the partial pressure of NO_2.

    II. The number of molecules of NO is the same as the number of molecules of NO_2.

    III. The number of nitrogen atoms is the same as the number of oxygen atoms.

    (A) I only

    (B) II only

    (C) III only

    (D) I and II only

    (E) I, II, and III

27. Carbon-14 decays by the emission of a particle to form nitrogen-14. What is the missing particle?

$$^{14}_{6}C \rightarrow {}^{14}_{7}N + ..?..$$

    (A) alpha

    (B) beta

    (C) gamma

    (D) positron

    (E) neutron

28. According to the Kinetic Molecular Theory, all of the following apply to a mixture of gases EXCEPT

(A) All gas molecules travel at the same speed.

(B) The collisions of the gas molecules are perfectly elastic.

(C) The forces of attraction between the gas molecules are negligibly small.

(D) The gas molecules exert pressure on the wall of the container of the system.

(E) Compared to the volume of the system, the absolute volume of the gas molecules is negligibly small.

29. Consider two identical flasks under identical conditions of temperature and pressure. One is filled with carbon monoxide, CO, while the other contains nitrogen, N_2. Which of the following characteristics is the same for each sample?

    I.   molar mass

    II. average molecular velocity

    III. average kinetic energy

(A) I only

(B) II only

(C) III only

(D) I and II only

(E) I, II, and III

30. Relatively high values for all the following physical properties are associated with strong intermolecular forces EXCEPT

(A) viscosity

(B) boiling point

(C) melting point

(D) vapor pressure

(E) critical temperature

Questions 31 and 32: Refer to the phase diagram for carbon dioxide given below.

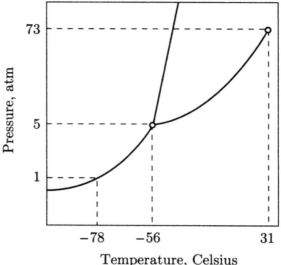

One sample of CO_2 is held at 10 atm and heated from $-60°C$ to $30°C$. A second sample of CO_2 is held at 1 atm and heated over the same temperature range.

31. Which of the following describes the expected properties of these samples?

  I.   The melting point for the first sample is observed at a temperature between $-78°C$ and $-56°C$.

  II. No melting occurs in the second example.

  III. The boiling point for the first sample is greater than the sublimation point for the second sample.

  (A) I only
  (B) III only
  (C) I and II only
  (D) II and III only
  (E) I, II and III

32. At 25°C and 1 atm, solid carbon dioxide does not melt but undergoes sublimation. Which information from the phase diagram supports the statement?

  (A) The pressure at the triple point is greater than 1 atm.
  (B) The boiling point at 1 atm is less than 25°C.
  (C) The pressure at the critical point is greater than 1 atm.
  (D) The temperature at the critical point is greater than 25°C.
  (E) The boiling point at 73 atm is greater than the boiling point at 5 atm.

33. The radionuclide, ^{32}P, is used to trace metabolic pathways. A 1.00 gram sample of ^{32}P decays to 0.50 grams in 14.3 days. Over the period of time of its use, the half-life of the nuclide

    (A) increases as rate of nuclear decay in grams per day increases.
    (B) increases as rate of nuclear decay in grams per day decreases.
    (C) remains unchanged as rate of nuclear decay in grams per day increases.
    (D) remains unchanged as rate of nuclear decay in grams per day decreases.
    (E) decreases as rate of nuclear decay in grams per day remains unchanged.

34. Which expression gives the density predicted for $PH_{3(g)}$ (molar mass: 34 g) at 740 mmHg and 20°C? (Assume ideal behavior.)

    (A) $\dfrac{34}{22.4 \times 740 \times 293}$

    (B) $\dfrac{34 \times 740 \times 293}{22.4}$

    (C) $\dfrac{34 \times 740 \times 273}{22.4 \times 760 \times 293}$

    (D) $\dfrac{34 \times 740 \times 293}{22.4 \times 760 \times 273}$

    (E) $\dfrac{34 \times 760 \times 273}{22.4 \times 740 \times 293}$

35. When excess HCl reacts with 0.250 mole of NH_3, a total of 11.0 kJ of heat energy is released. What is the value of ΔH in kJ per mol NH_3 consumed?

    (A) −44.0 kJ
    (B) −11.0 kJ
    (C) −2.75 kJ
    (D) +11.0 kJ
    (E) +44.0 kJ

Questions 36 and 37: Information about NaCl at 298 K is given below:

$$NaCl_{(s)} \quad \Delta H_f^\circ = -410 \text{ kJ mol}^{-1} \qquad \Delta G_f^\circ = -384 \text{ kJ mol}^{-1}$$

$$NaCl_{(aq)} \quad \Delta H_f^\circ = -407 \text{ kJ mol}^{-1} \qquad \Delta G_f^\circ = -393 \text{ kJ mol}^{-1}$$

36. According to the information above, which characteristics apply to the dissolving process for $NaCl_{(s)}$ at 298 K?

    (A) spontaneous and endothermic
    (B) spontaneous and exothermic
    (C) not spontaneous and endothermic
    (D) not spontaneous and exothermic
    (E) equilibrated at standard conditions

37. If the temperature is increased to 50°C, the value of ΔG for the dissolving process

    (A) decreases as the value of $T\Delta S$ increases
    (B) decreases as the value of $T\Delta S$ decreases
    (C) decreases as the value of $T\Delta S$ remains the same
    (D) remains the same as the value of $T\Delta S$ increases
    (E) remains the same as the value of $T\Delta S$ decreases

---

38. Which 0.1 m aqueous solution has the lowest freezing point at 1 atm?

    (A) $CaCl_2$
    (B) $NaNO_3$
    (C) CH_3OH
    (D) $C_6H_{12}O_6$
    (E) $(NH_4)_3PO_4$

39. What mass of H_3PO_4 should be added to 800 grams of water to produce a solution that is 20% H_3PO_4 by mass?

    (A)   150 g
    (B)   200 g
    (C)   900 g
    (D) 1,800 g
    (E) 2,400 g

40. A solution of HCl of unknown concentration is analyzed using a solution of $Ba(OH)_2$. Exact neutralization occurs after 15. mL of 0.30 M $Ba(OH)_2$ has been added to 30. mL of HCl. What is the concentration of the unknown HCl solution?

(A) 0.075 M
(B) 0.10 M
(C) 0.15 M
(D) 0.17 M
(E) 0.30 M

41. Many sedimentary rocks include the carbonate ion, a remnant of the shells of ancient sea creatures. A geologist tests for the presence of carbonate ions by placing several drops of concentrated acid on the rock, then waits to see if bubbles occur. Which property accounts for this behavior?

(A) Calcium carbonate is an ionic solid.
(B) Carbonate ion is a strong proton acceptor.
(C) Acidic groundwater acts as a dehydrating agent.
(D) In acid solution, carbonate reacts with oxygen.
(E) In acid solution, carbonate is reduced to oxalate.

42.
$$C_2H_5OH_{(\ell)} + 3O_{2(g)} \rightarrow 2CO_{2(g)} + 3H_2O_{(\ell)}$$

Which expression gives the volume of oxygen measured at 1 atm and 273 K that is consumed when 50 g C_2H_5OH burns in excess oxygen according to the equation above?

(A) $50 \times 3 \times 22.4$

(B) $\dfrac{50 \times 22.4}{3 \times 46}$

(C) $\dfrac{50 \times 3}{46 \times 22.4}$

(D) $\dfrac{50 \times 22.4}{46}$

(E) $\dfrac{50 \times 3 \times 22.4}{46}$

43. According to the Bronsted-Lowry theory of acid-base behavior, HPO_4^{2-} is classified as amphiprotic because

(A) it reacts with OH^-
(B) it reacts with H_3O^+
(C) its ionic charge is 2^-
(D) its parent molecule is polyprotic
(E) it can accept and donate protons

44.
$$2Al_{(s)} + 3Cl_{2(g)} \rightarrow 2AlCl_{3(s)}$$

A mixture containing 0.40 mol Al in contact with 0.75 mol Cl_2 is ignited in a closed container. Assuming complete reaction, which gives the maximum quantity of $AlCl_3$ produced and the quantity of unconsumed reactant?

(A) 0.40 mol $AlCl_3$ and 0.15 mol Al unreacted

(B) 0.40 mol $AlCl_3$ and 0.15 mol Cl_2 unreacted

(C) 0.40 mol $AlCl_3$ and 0.35 mol Cl_2 unreacted

(D) 0.50 mol $AlCl_3$ and 0.10 mol Al unreacted

(E) 0.50 mol $AlCl_3$ and 0.15 mol Cl_2 unreacted

45. How many of the halide salts listed below have percent halogen by mass that is less than 50%?

$$AgI \quad BaF_2 \quad CaCl_2 \quad FeBr_2 \quad PbCl_2$$

(A) one
(B) two
(C) three
(D) four
(E) five

46. What is the final concentration of chloride ion when 50 mL of 0.20 M $MgCl_{2(aq)}$ is mixed with 100 mL of 0.10 M $KCl_{(aq)}$?

(A) 0.10 M
(B) 0.15 M
(C) 0.20 M
(D) 0.30 M
(E) 0.40 M

47. At 50°C, the vapor pressures of carbon tetrachloride and chloroform are approximately 300 mmHg and 500 mmHg, respectively. Which expression gives the vapor pressure, in mmHg, of a solution that is 0.20 mole fraction carbon tetrachloride in chloroform?

(A) $0.20 \times (300 + 500)$

(B) $0.20 \times (500 - 300)$

(C) $\dfrac{0.20 \times 300}{0.80 \times 500}$

(D) $\dfrac{0.20 \times 500}{0.80 \times 300}$

(E) $(0.20 \times 300) + (0.80 \times 500)$

48. Which is the best net ionic equation for the reaction between a solution of sodium sulfite and excess phosphoric acid solution?

(A) $SO_3^{2-} + 2H^+ \rightarrow H_2O + SO_2$

(B) $3Na^+ + PO_4^{3-} \rightarrow Na_3(PO_4)_2$

(C) $Na_2SO_3 + 2H^+ \rightarrow Ca^{2+} + H_2O + SO_2$

(D) $3SO_3^{2-} + 2H_3PO_4 \rightarrow 3H_2O + 3SO_2 + 2PO_4^{3-}$

(E) $3Na_2SO_3 + 2H_3PO_4 \rightarrow 2Na_3PO_4 + 3H_2O + 3SO_2$

49. $\ldots H_2S_{(aq)} + \ldots IO_3^-{}_{(aq)} \rightarrow \ldots I_{2(aq)} + \ldots SO_3^{2-}{}_{(aq)} + \ldots H_2O + \ldots H^+{}_{(aq)}$

When the reaction above is balanced and all of the coefficients are reduced to their lowest whole-number values, how many moles of electrons are transferred?

(A) 4

(B) 6

(C) 8

(D) 12

(E) 30

50. Which of the following is least likely to behave as a Lewis base?

(A) H_2O

(B) Cl^-

(C) BF_3

(D) OH^-

(E) NH_3

51.

Acid	pK$_a$
H_3PO_4	2.1
$H_2PO_4^-$	7.1
HPO_4^{2-}	12.3

A buffer prepared by mixing equimolar solutions of H_3PO_4 and NaH_2PO_4 and Na_2HPO_4 will have a pH closest to

(A) 2

(B) 4

(C) 7

(D) 10

(E) 12

52. Consider the equilibrium

$$H_{2(g)} + I_{2(g)} \rightleftharpoons 2HI_{(g)}$$

In a closed system, the initial partial pressure of hydrogen is 1.25 atm and the initial partial pressure of iodine is 1.75 atm. At equilibrium, the partial pressure of hydrogen is 1.00 atm. Which expression gives the value of K_p for this system?

(A) $\dfrac{(0.50)^2}{(1.00)(1.75)}$

(B) $\dfrac{(0.50)^2}{(1.50)^2}$

(C) $\dfrac{(0.25)^2}{(1.25)(1.75)}$

(D) $\dfrac{(0.50)^2}{(1.00)(1.50)}$

(E) $\dfrac{(0.25)^2}{(1.00)(1.50)}$

53. The pH of 0.10 M solution of $NaHSO_4$ is closest to

(A) 1
(B) 3
(C) 7
(D) 11
(E) 14

54. $$H_3O^+ + NH_2^- \rightleftharpoons NH_3 + H_2O$$

In the proton transfer shown in the equation above, all of the following are correct EXCEPT

(A) NH_3 acts as a Bronsted acid.
(B) The value of K is greater than 1.
(C) The H-NH_2 bond is weaker than the H-H_2O bond.
(D) One of the conjugate acid/base pairs is H_3O^+ and H_2O.
(E) NH_2^- and H_2O have the same number of unshared electron pairs.

55. Which value is closest to the molar solubility of $Ca(OH)_2$?

$$Ca(OH)_{2(s)} \rightleftharpoons Ca^{2+}{}_{(aq)} + 2\, OH^-{}_{(aq)} \qquad K_{sp} = 8 \times 10^{-6}$$

(A) 1×10^{-2}

(B) 2×10^{-2}

(C) 3×10^{-3}

(D) 2×10^{-6}

(E) 8×10^{-6}

56. Approximately what minimum quantity of charge, measured in coulombs, is needed to produce 0.15 moles of chromium metal in an electrolytic cell that contains 0.75 M solution of $Cr(NO_3)_3$?

(A) 5,000 coulombs

(B) 15,000 coulombs

(C) 25,000 coulombs

(D) 45,000 coulombs

(E) 225,000 coulombs

57. $$Ag^+ + e^- \rightarrow Ag^0 \qquad E^\circ = 0.80 \text{ volts}$$

$$Cd^{2+} + 2e^- \rightarrow Cd^0 \qquad E^\circ = -0.40 \text{ volts}$$

Which expression gives the voltage produced in a standard chemical cell using the half-reactions above?

(A) $0.80 + 0.40$

(B) $0.80 - 0.40$

(C) $0.40 - 0.80$

(D) $0.80 + (2 \times (-0.40))$

(E) $(2 \times (-0.80)) + 0.40$

58. $$Cu_{(s)} + 2Ag^+{}_{(aq)} \rightarrow Cu^{2+}{}_{(aq)} + 2Ag_{(s)} \qquad K_{eq} = 3.75 \times 10^{15}$$

Which describes the standard voltage and standard free energy change for this reaction?

(A) $E^\circ < 0;\ \Delta G^\circ > 0$

(B) $E^\circ > 0;\ \Delta G^\circ < 0$

(C) $E^\circ > 0;\ \Delta G^\circ > 0$

(D) $E^\circ < 0;\ \Delta G^\circ < 0$

(E) $E^\circ = \Delta G^\circ = 0$

59.

$$Zn^{2+} + 2e^- \rightarrow Zn^0 \qquad E° = -0.76 \text{ volts}$$

$$Al^{3+} + 3e^- \rightarrow Al^0 \qquad E° = -1.66 \text{ volts}$$

When half-cells based on the half-reactions above are used to construct an electrochemical cell, the overall reaction becomes

$$2Al + 3Zn^{2+} \rightarrow 3Zn + 2Al^{3+}$$

Which gives the effect on the cell voltage when a standard aluminum-zinc cell is changed to a cell in which all ions have 0.01 M concentration?

(A) No change is observed.
(B) The voltage decreases by about 0.02 volts.
(C) The voltage decreases by about 0.12 volts.
(D) The voltage increases by about 0.02 volts.
(E) The voltage increases by about 0.12 volts.

60. Consider a system at equilibrium according to the equation

$$N_{2(g)} + 3H_{2(g)} \rightleftharpoons 2NH_{3(g)}$$

If $Ar_{(g)}$ is added to such an equilibrium system at constant volume and temperature, the total pressure

(A) increases and the number of NH_3 molecules present increases
(B) decreases and the number of NH_3 molecules present increases
(C) remains the same and the number of NH_3 molecules present increases
(D) remains the same and the number of NH_3 molecules present decreases
(E) increases and the number of NH_3 molecules present remains the same

61. Which is a correct comparison of the characteristics of a catalyzed reaction to the corresponding characteristics of the same reaction without a catalyst present?

   I.   Their energies of activation are the same.
   II. Their enthalpies of reaction are the same.
   III. Their free energies of reaction are the same.

(A) I only
(B) II only
(C) II and III only
(D) I and III only
(E) I, II and III

62. Consider a system at equilibrium based on the equation

$$2NO_{(g)} + Cl_{2(g)} \rightleftharpoons 2NOCl_{(g)}$$

Which is a correct statement about this equilibrium system at constant temperature?

I.   The concentration of NOCl is constant.

II. The rate of loss of Cl_2 is equal to the rate of formation of NOCl.

III. The rate of formation of NO is equal to the rate of loss of NOCl.

(A) I only
(B) III only
(C) I and III only
(D) II and III only
(E) I, II, and III

Questions 63 and 64: The reaction between chlorine and chloroform in the gas phase is known to proceed according to the mechanism below

$$Cl_{2(g)} \rightleftharpoons 2Cl_{(g)} \qquad \text{fast equilibrium}$$
$$Cl_{(g)} + CHCl_{3(g)} \rightarrow HCl_{(g)} + CCl_{3(g)} \qquad \text{slow}$$
$$CCl_{3(g)} + Cl_{(g)} \rightarrow CCl_{4(g)} \qquad \text{fast}$$

63. According to this mechanism, what is the overall reaction?

(A) $Cl_{2(g)} \rightarrow CCl_{4(g)}$

(B) $CHCl_{3(g)} + Cl_{2(g)} \rightarrow HCl_{(g)} + CCl_{4(g)}$

(C) $Cl_{(g)} + CHCl_{3(g)} \rightarrow H_{(g)} + CCl_{4(g)}$

(D) $Cl_{(g)} + CHCl_{3(g)} \rightarrow HCl_{(g)} + CCl_{3(g)}$

(E) $2CHCl_{3(g)} + Cl_{2(g)} \rightarrow 2HCl_{(g)} + CCl_{4(g)} + CCl_{3(g)}$

64. According to this mechanism, what is the rate law?

(A) $RATE = k[Cl_2]^2$

(B) $RATE = k[Cl_2]^2[CHCl_3]$

(C) $RATE = k[Cl_2]^{\frac{1}{2}}[CHCl_3]$

(D) $RATE = k\dfrac{[CHCl_3]}{[Cl_2]^{\frac{1}{2}}}$

(E) $RATE = k\dfrac{[CHCl_3]}{[Cl]^2}$

65. Zinc oxalate, ZnC_2O_4, and lead fluoride, PbF_2, have the same K_{sp} value at 25°C, 2.7×10^{-8}. Which can be concluded from this information?

    (A) Their absolute entropies in J mol^{-1} K^{-1} are equal.

    (B) The molar solubility of lead fluoride is greater.

    (C) Their K_{sp} values are equal at all temperatures.

    (D) The heat of solution in kJ mol^{-1} for zinc oxalate is greater.

    (E) Their solubilities in grams of solute per liter of solution are equal.

66. Consider the equilibrium represented by the equation

$$H_{2(g)} + F_{2(g)} \rightleftharpoons 2HF_{(g)}$$

    Which applies to this system at equilibrium?

    I. $K_p = K_c$

    II. $\Delta G = $ zero

    III. $Rate_{forward} = Rate_{reverse}$

    (A) I only

    (B) II only

    (C) I and II only

    (D) II and III only

    (E) I, II and III

67. The reaction between H_2 and NO occurs according to the equation

$$2H_{2(g)} + 2NO_{(g)} \rightarrow 2H_2O_{(g)} + N_{2(g)}$$

    Six trials of the reaction were carried out. The initial rate of change of pressure for each trial was measured and recorded.

	Initial Pressure (atm)		Initial Rate
Trial	P_{NO}	P_{H_2}	Δatm min^{-1}
I	0.50	0.09	0.025
II	0.50	0.18	0.050
III	0.50	0.27	0.075
IV	0.09	0.80	0.0063
V	0.18	0.80	0.025
VI	0.27	0.80	0.056

    Based on these results, what is the rate law for this reaction?

    (A) $RATE = k(P_{NO})^0(P_{H_2})^2$

    (B) $RATE = k(P_{NO})^1(P_{H_2})^2$

    (C) $RATE = k(P_{NO})^2(P_{H_2})^0$

    (D) $RATE = k(P_{NO})^2(P_{H_2})^1$

    (E) $RATE = k(P_{NO})^2(P_{H_2})^2$

68. A worker was assigned to prepare a 1.0 molal aqueous solution of ethanol, C_2H_5OH (molar mass 46, density 0.79 g mL^{-1}), using 0.50 kg water. The worker used a volumetric flask to measure 500 mL of water which was poured into a 1-liter beaker. The worker then used a buret to measure 23 mL of ethanol which was added to the contents of the beaker. Which describes the resulting solution?

(A) The solution was correctly prepared to the assigned concentration.

(B) The concentration of the solution will be greater than 1.0 m because too much water was used.

(C) The concentration of the solution will be less than 1.0 m because too much water was used.

(D) The concentration of the solution will be greater than 1.0 m because too much ethanol was used.

(E) The concentration of the solution will be less than 1.0 m because too little ethanol was used.

69. The combustion of ethyne occurs according to the equation

$$2C_2H_{2(g)} + 5O_{2(g)} \rightarrow 4CO_{2(g)} + 2H_2O_{(\ell)} \qquad \Delta H^\circ = ? \text{ kJ}$$

$$\Delta H^\circ_f \text{ for } H_2O_{(\ell)} = -300 \text{ kJ mol}^{-1}$$

$$\Delta H^\circ_f \text{ for } CO_{2(g)} = -400 \text{ kJ mol}^{-1}$$

$$\Delta H^\circ_f \text{ for } C_2H_{(g)} = 200 \text{ kJ mol}^{-1}$$

The approximate standard heats of formation for water, carbon dioxide and ethyne are given above. Based on this information, which of the following values is closest to ΔH° in kJ mol^{-1} for this reaction?

(A) -900 kJ mol^{-1}
(B) -1300 kJ mol^{-1}
(C) -1800 kJ mol^{-1}
(D) -2300 kJ mol^{-1}
(E) -2600 kJ mol^{-1}

70. Which is a common ingredient of agricultural fertilizers?

(A) copper sulfate
(B) calcium sulfide
(C) barium phosphate
(D) aluminum chloride
(E) ammonium nitrate

71. $Ba(OH)_2 \cdot 8H_2O_{(s)} + 2NH_4SCN_{(s)} \rightarrow Ba(SCN)_{2(aq)} + 2NH_{3(aq)} + 10H_2O_{(\ell)}$
When solid samples of barium hydroxide and ammonium thiocyanate are mixed in a test tube, the outside of the test tube becomes noticeably cool to the touch. Which gives the correct signs for the thermodynamic parameters for such a spontaneous endothermic dissolving process?

	ΔG_{soln}	ΔH_{soln}	ΔS_{soln}
(A)	+	+	+
(B)	+	+	−
(C)	−	−	+
(D)	−	+	+
(E)	−	−	−

72. An experiment to determine the percent of water in a hydrated salt was carried out by heating the hydrated salt over an intense flame. During the heating process, and unknown to the worker, some of the material being heated was accidentally spilled. Which describes the effect on the reported results of the experiment?

    (A) Both the mass of the hydrated salt and the percent water will be reported too larger.
    (B) The mass of the hydrated salt will be reported correctly and the percent water will be reported too small.
    (C) The mass of the hydrated salt will be reported correctly and the percent water will be reported too large.
    (D) The mass of the hydrated salt will be reported too large and the percent water will be reported too small.
    (E) The mass of the hydrated salt will be reported too small and the percent water will be reported too large.

73. Which products are formed when crystals of $FeSO_4$ are dissolved in an acidified solution of $KMnO_4$?

    (A) Mn^{2+} and Fe^{3+}
    (B) Mn^{2+} and Fe
    (C) Mn^{2+} and SO_2
    (D) MnO_2 and Fe
    (E) MnO_2 and SO_2

74. Which element is most likely to be found in compounds taken internally to be used for x-ray diagnosis of disorders of the intestinal tract?

    (A) barium
    (B) lead
    (C) lithium
    (D) magnesium
    (E) zinc

75. Ceramics are materials known for their strength, brittleness, and resistance to heat and chemical corrosion. The inclusion of which of the following is responsible for these characteristics?

    (A) silicates
    (B) organometallics
    (C) high temperature metals
    (D) polymers
    (E) semiconductors

Section II

Section II - Free Response   Total Time – 90 Minutes
(Multiple-Choice Questions are found in Section I.)

Part A: Question 76
and
Question 77 or Question 78
Time: 40 minutes

Access to calculators, Periodic Table, lists of standard reduction potentials, and
Equations and Constants

(2004 Examination directions) Clearly show the method used and the steps involved in arriving at your answers. It is to your advantage to do this, because you may obtain partial credit if you do and you will receive little or no credit if you do not. Attention should be paid to significant figures. Be sure to write all your answers to the questions on the lined pages following each question in the booklet with the pink cover. Do not write your answers on the green insert.

Answer question 76 below. The Section II score weighting for this question is 20 percent.

76.   Answer the following questions concerning the homogeneous gaseous equilibrium system represented by the equation below.

$$2\,HI_{(g)} \rightleftharpoons H_{2(g)} + I_{2(g)}$$

At a certain temperature, $HI_{(g)}$ is inserted into a tank until the pressure is 1.00 atm. After the system reaches equilibrium, the partial pressure of the HI is 0.80 atm.

(a)   Write the equilibrium expression, K_p, for this reaction.

(b)   Calculate the partial pressure of $H_{2(g)}$ and $I_{2(g)}$ at equilibrium.

(c)   Calculate the value of K_p for this reaction.

(d)   Calculate the value of K_c for this reaction.

(e)   What additional information would you need to be able to calculate the number of moles of HI present in the tank at equilibrium?

Answer either question 77 or question 78 below.

(2004 examination directions) Only one of these two questions will be graded. If you start both questions, be sure to cross out the question you do not want graded.

The Section II score weighting for the question that you choose is 20 percent.

77. Answer all five questions below related to an aqueous solution of the weak acid, HF. Hydrogen fluoride, HF, a weak acid, dissolves in water to form hydrofluoric acid according to the following equation.

$$HF_{(aq)} \rightleftharpoons H^+_{(aq)} + F^-_{(aq)}$$

   (a) Available in the laboratory are 40.0 grams of HF and 2.00 kg water. Determine the mass of each substance which is needed to make the maximum quantity of 0.100 molal HF solution.

   (b) Determine the molality of F^- ions and the molality of HF molecules in the solution prepared as directed in part (a).

   (c) Determine the percent ionization of HF in this solution.

   (d) The freezing point of the 0.100 m solution of HF in water is $-0.201°C$. Determine the apparent molality of all dissolved particles.

   (e) Determine the number of moles of HF molecules in the solution described in part (a).

78. Use the information in this table of standard heats of formation and standard entropies to answer all four questions below related to the combustion of acetic acid.

	$\Delta H_f°$ kJ mol^{-1}	S° J mol^{-1} K^{-1}
$CO_{2(g)}$	-393.5	213.6
$C_{(s)}$	0	5.69
$H_{2(g)}$	0	130.58
$H_2O_{(\ell)}$	-285.83	69.96
$CO_{2(g)}$	0	205.0
$CH_3COOH_{(\ell)}$	?	159.81

All values taken at 298 K.

At 298 K, the standard heat (enthalpy) of combustion, $\Delta H°_{comb}$, of acetic acid, $CH_3COOH_{(\ell)}$, is -874.5 kJ mol^{-1}.

   (a) Write the balanced equation for the complete combustion of acetic acid, $CH_3COOH_{(\ell)}$, in pure oxygen.

   (b) Calculate the standard entropy change, $\Delta S°$, for this combustion reaction at 298 K. Specify units.

   (c) Calculate the standard enthalpy of formation, $\Delta H_f°$, for acetic acid at 298 K. Specify units.

   (d) Calculate the standard Gibbs free energy change, $\Delta G°$, for this combustion reaction at 298 K. Specify units.

Part B: Questions 79, 80, 81 and
Question 82 or Question 83
Time: 50 minutes

Access to Periodic Table, lists of standard reduction potentials
and *Equations and Constants*
No access to calculators

Answer question 79 below: The Section II score weighting for this question is 15 percent.

79. (2004 Examination directions) Write the formulas to show the reactants and products for FIVE of the laboratory situations described below. Answers to more than five choices will not be graded. In all cases a reaction occurs. Assume that solutions are aqueous unless otherwise indicated. Represent substances in a solution as ions if the substances are extensively ionized. Omit formulas for any ions or molecules that are unchanged by the reaction. You need not balance the equations.

(a) Solutions of mercury(I) nitrate and hydrochloric acid are mixed.

(b) Sulfur dioxide gas is bubbled through water.

(c) A bar of manganese metal is added to a solution of zinc nitrate.

(d) Excess concentrated ammonia is added to a solution of copper(II) nitrate.

(e) A quantity of 1-propanol is burned in oxygen.

(f) Solutions of iron(II) chloride and gold(III) nitrate are mixed.

(g) Solutions of ammonium chloride and sodium fluoride are mixed.

(h) Solutions of potassium sulfate and strontium bromide are mixed.

(2004 Examination directions) Your responses to the rest of the questions in this part of the examination will be graded on the basis of the accuracy and relevance of the information cited. Explanations should be clear and well organized. Examples and equations may be included in your responses where appropriate. Specific answers are preferable to broad, diffuse responses.

(2004 examination directions) **Answer both Question 80 and Question 81 below.** Both questions will be graded.

The Section II score weighting for these questions is 30 percent (15 percent each).

80. Use principles of chemical bonding with appropriate Lewis dot structures to answer the four questions below.

    (a) Draw the Lewis electron-dot structure and predict the shape of the two halide species below.

    $$SF_4, \; ICl_4^-$$

    (b) Use the valence shell electron pair repulsion (VSEPR) model to explain the geometry of the electron pair distribution for each of the species in part (a).

    (c) Use Lewis electron-dot structures to show how each of the molecules below is an illustration of an exception to the octet rule.

    electron deficient molecule: BF_3
    (fewer than eight electrons)

    expanded octet: ICl_3

81. Answer all four questions below about the experimental determination of the concentration of an unknown solution of acetic acid, CH_3COOH.

A dilute solution of NaOH of known concentration can be used to determine the unknown concentration of a solution of CH_3COOH. The solution of NaOH can be titrated against a measured quantity of the unknown CH_3COOH solution. The endpoint can be determined by using a few drops an appropriate indicator added to the acid or by monitoring the pH of the reaction mixture.

Use the axes below for the answers to part (a) and part (b). Note that a line for pH = 7 has been given.

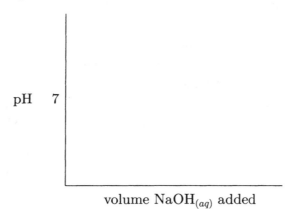

volume $NaOH_{(aq)}$ added

(a) On the axes above, draw a curve labeled (A) to show how $[H^+]$ changes in the reaction mixture as $NaOH_{(aq)}$ is added to $CH_3COOH_{(aq)}$. Explain.

(b) Explain the changes $[CH_3COO^-]$ as this reaction proceeds.

(c) On the same axes used for part (a), draw a new curve labeled (C) to show how the results would be different if the volume of acid neutralized were increased. Indicate on your new graph where neutralization is observed. Explain.

(d) Of the indicators described below, choose the one that would be most <u>unsatisfactory</u> for detection of the neutralization for this reaction. Explain.

Indicator	pH range for color change
methyl red	$4.8 - 6.0$
bromthymol blue	$6.0 - 7.6$
thymol blue	$8.0 - 9.6$

(2004 examination directions) **Answer either question 82 or question 83 below.** Only one of these two questions will be graded. If you start both questions, be sure to cross out the question you do not want graded. The Section II score weighting for the question that you choose is 15 percent.

82. Answer the four questions below about the chemical reaction.

$$H_{2(g)} + I_{2(g)} \rightarrow 2HI_{(g)}$$

For the exothermic reaction represented above, carried out at 298 K, the Rate Law is RATE $= k[H_2][I_2]$. Four each of the four changes, (a) through (d), below, predict the effect of that change on the initial rate of the reaction. Explain each in terms of the collision theory of reaction kinetics.

(a) **Addition of $H_{2(g)}$ at constant temperature and volume.** Include an energy distribution diagram showing number of molecules as function of energy before and after the addition of $H_{2(g)}$.

(b) **Increase in volume of reaction vessel at constant temperature.** Include an energy distribution diagram showing $[H_2]$ as function of energy before and after the increase in volume.

(c) **Addition of catalyst.** Include a diagram where potential energy is plotted on the reaction coordinate. Indicate potential energy for the activated complex with and without the catalyst. Label the curve that represents the catalyzed reaction.

(d) **Increase in temperature.** Include an energy distribution diagram showing number of molecules as function of energy before and after the increase in temperature. Locate E_a. Use features of your diagram to help explain why increase in temperature causes an increase in rate of reaction.

83. Answer all four questions below related to properties of the alkali metal, potassium.

(a) The atomic radius of potassium is greater than the atomic radius of zinc. Explain.

(b) The radius of the K^+ ion is smaller than the radius of the K^0 atom. Explain.

(c) The ionic radius of Cl^- is greater than the ionic radius of K^+. Explain.

(d) The second ionization energy of potassium is greater than the second ionization energy of calcium. Explain.

SAMPLE EXAMINATION III

Section I – Multiple Choice

Questions 1-5. The set of lettered choices below, a list of frequently used ions, refers to the set of numbered statements immediately following it. Select the one lettered choice that is most closely associated with each numbered statement. Each lettered choice can be used once, more than once or not at all.

(A) Cl^-

(B) Br^-

(C) Ca^{2+}

(D) NH_4^+

(E) OH^-

Assume that you have an "unknown" that is an aqueous solution of a substance which contains one or more of the ions listed above. Which ion must be ABSENT based upon each of the following observations of the "unknown"?

1. The pH of the unknown solution decreases immediately and rapidly as dilute $HNO_{3(aq)}$ is added dropwise.

2. No ammonia odor is detected when dilute NaOH solution is added to a sample of the unknown solution and warmed.

3. No white precipitate is observed when Na_2CO_3 solution is added to a sample of the unknown solution.

4. No reddish-brown precipitate is observed when $FeCl_3$ solution is added to a sample of the unknown solution.

5. No yellow precipitate is observed when $AgNO_3$ solution is added to a sample of the unknown solution.

---

6. When Fe^0 changes to Fe^{3+} the number of occupied orbitals changes from
    (A) 15 to 14
    (B) 15 to 12
    (C) 15 to 9
    (D) 13 to 12
    (E) 13 to 11

245

7. Which formula matches the name, pentaamminechlorocobalt(III) chloride?

    (A) $[Co(NH_3)_5]Cl_3$
    (B) $[Co(NH_3)_5Cl]Cl_3$
    (C) $[Co(NH_3)_5Cl]Cl_2$
    (D) $[Co(NH_3)Cl]_5Cl_{10}$
    (E) $[Co(NH_3Cl)_5Cl]Cl_2$

8. In the dissolving of solid $CaCl_2$ in water to form an aqueous solution, bonds in $CaCl_2$ between

    (A) ions are broken as ion-dipole bonds form
    (B) ions are broken as covalent bonds form
    (C) atoms are broken as covalent bonds form
    (D) molecules are broken as ion-dipole bonds form
    (E) molecules are broken as dipole-dipole bonds form

Questions 9-13: These questions refer to the electrochemical cell shown below.

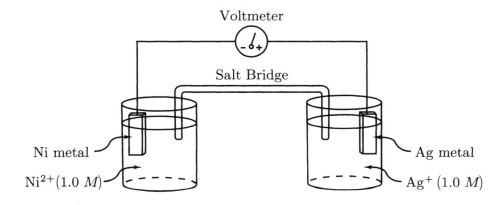

The cell above is constructed using the half reactions below.

$$Ag^+ + e^- \rightarrow Ag^0 \qquad E° = 0.80 \text{ V}$$
$$Ni^{2+} + 2e^- \rightarrow Ni^0 \qquad E° = -0.23 \text{ V}$$

9. Which expression gives the value of $\Delta G°$ in joules for this cell?

    (A) $-2 \times 96{,}500 \times 1.03$

    (B) $-8.31 \times 298 \times \ln 1.03$

    (C) $-2 \times 96{,}500 \times \ln 1.03$

    (D) $\dfrac{-2 \times 96{,}500 \times 1.03}{8.31}$

    (E) $-8.31 \times 298 \times \ln \dfrac{0.80}{0.23}$

Questions 10-13: The set of lettered choices is a list of possible observations about the voltage of the electrochemical cell shown above. Changes in the cell are described in statements 10 through 13. Select the one lettered choice that is most closely related to each statement.

    (A) Voltage increases.
    (B) Voltage decreases but remains greater than zero.
    (C) Voltage decreases to zero and remains zero.
    (D) Voltage decreases to a value less than zero.
    (E) No change in voltage occurs.

Which of the above is observed for each of the following changes in the system?

10. Another 100 mL of 1.0 M $Ni(NO_3)_2$ solution is added to the Ni^{2+} solution.

11. The surface area of the silver strip is increased.

12. The salt bridge is replaced by a platinum wire.

13. A small quantity of $AgNO_{3(s)}$ is added to the Ag^+ solution.

---

14. Consider the following quantum numbers for two different electrons in a ground state atom of phosphorus. Which is a correct comparison of these electrons?

$$3, 1, 1, -\tfrac{1}{2} \qquad 3, 1, 0, -\tfrac{1}{2}$$

    (A) These electrons have the same energy and occupy different orbitals.
    (B) These electrons have the same energy and occupy the same orbital.
    (C) These electrons have different energies and occupy different orbitals.
    (D) These electrons have different energies and occupy different energy sublevels.
    (E) These electrons have different energies and occupy the same energy sublevels.

15. Both chlorine and bromine exist as two naturally-occurring isotopes, distributed as shown below. The "percent natural occurrence" is based on distribution of isotopes. Chlorine reacts with bromine to form ClBr.

isotope	percent natural occurrence
chlorine – 35	76%
chlorine – 37	24%
bromine – 79	51%
bromine – 81	49%

Thus there are three different possible molar masses for ClBr: 114, 116, and 118. Which of the following gives the approximate fractional distribution in naturally-occurring compounds of the three possible molar masses?

molar masses

	114	116	118
(A)	$\frac{1}{4}$	$\frac{1}{2}$	$\frac{1}{4}$
(B)	$\frac{1}{3}$	$\frac{1}{3}$	$\frac{1}{3}$
(C)	$\frac{3}{8}$	$\frac{1}{2}$	$\frac{1}{8}$
(D)	$\frac{1}{2}$	$\frac{1}{4}$	$\frac{1}{4}$
(E)	$\frac{1}{2}$	$\frac{3}{8}$	$\frac{1}{8}$

16. Which is a correct statement of a trend within a group of elements on the Periodic Table as atomic number increases?

    I.   The number of valence electrons increases.

    II.   The radius of the most common ion of each element increases.

    III.   The ionization energy increases.

(A) I only

(B) II only

(C) III only

(D) I and II only

(E) I, II, and III

17. Consider the oxides, XO_2 and YO_2, where X and Y are nonmetals with nearly equal atomic mass. The molecular shape of XO_2 is linear; that of YO_2 is bent (V-shaped). Which gives a correct comparison of their properties?

    I.   The critical temperature of YO_2 is higher than that of XO_2.

    II.   The normal boiling point of YO_2 is higher than that of XO_2.

    III. The molar heat of vaporization of YO_2 is greater than that of XO_2.

(A) I only

(B) III only

(C) I and II only

(D) II and III only

(E) I, II, and III

18. In general, the melting points of ionic solids are higher than the melting points of molecular solids. Which accounts for this difference?

(A) Attractions between particles with opposite charge are stronger than covalent bonds.

(B) Repulsions between ions with the same charge are negligibly small.

(C) Delocalized electron clouds in molecular solids are produced by the formation of *pi* bonds.

(D) Attractions between ions with opposite charge are stronger than intermolecular forces.

(E) The distance between oppositely charged ions is less than the corresponding distance between bonded atoms.

19. Compared to a molecule of $C_2H_4Cl_2$, a molecule of $C_2H_2Cl_2$ contains

    I.   two fewer *sigma* bonds

    II.   one more *pi* bond

    III. two fewer shared electron pairs

(A) I only

(B) II only

(C) I and II only

(D) II and III only

(E) I, II, and III

20. For the compound PF_3, which describes the molecular shape and the distribution of valence shell electron pairs on the central atom?

	distribution of valence shell electron pairs	**molecular shape**
(A)	trigonal bipyramidal	trigonal planar
(B)	trigonal bipyramidal	T-shaped
(C)	tetrahedral	trigonal pyramidal
(D)	tetrahedral	tetrahedral
(E)	trigonal bipyramidal	tetrahedral

21. In which one of the following five compounds is the number of elements different from the number of elements in the other four?

(A) potassium permanganate
(B) potassium acetate
(C) potassium oxalate
(D) potassium perchlorate
(E) potassium selenate

22. Which will cause an increase in the mean free path for molecules in a sample of gas?

(A) increase pressure at constant volume
(B) increase temperature at constant volume
(C) increase density at constant temperature
(D) increase temperature at constant pressure
(E) increase pressure at constant temperature

23. The density of an unknown gas is determined to be 1.50 g L^{-1}. At the same conditions of temperature and pressure, the density of oxygen is determined to be 1.25 g L^{-1}. Which expression gives the molar mass of the unknown gas?

(A) $\dfrac{32 \times 1.25}{1.50}$

(B) $\dfrac{32 \times 1.50}{1.25}$

(C) $\dfrac{32 \times 1.25}{2 \times 1.50}$

(D) $\dfrac{32 \times (1.25)^2}{(1.50)^2}$

(E) $\dfrac{32 \times (1.50)^2}{(1.25)^2}$

24. Which occurs during the vaporization of a liquid at its normal boiling point?

    (A) The potential energy of the molecules decreases.
    (B) The average kinetic energy of the molecules increases.
    (C) The vapor pressure of the liquid decreases as the liquid is converted to vapor.
    (D) Attractive forces between atoms in a molecule are overcome as translational motion increases.
    (E) Attractive forces between molecules in a liquid are overcome as translational motion increases.

25. Which is an example of a p-type semiconductor; that is, a semiconductor in which a transport of charge is produced by moving spaces that accommodate valence electrons?

    (A) arsenic with some silicon added
    (B) germanium with some silicon added
    (C) silicon with some gallium added
    (D) silicon with some germanium added
    (E) germanium with some arsenic added

26. How many moles of water must be added to 20 moles of ethanol in order to prepare a solution that is 0.25 mole fraction in ethanol?

    (A)   5.0
    (B)   15
    (C)   60.
    (D)   80.
    (E)   75

27. At 298 K, as the salt MX dissolves spontaneously to form an aqueous solution, ΔS and ΔH are positive. Which describes the value of ΔG and the absolute values of its components, $T\Delta S$ and ΔH?

    (A) $\Delta G < 0$; $|T\Delta S| > |\Delta H|$
    (B) $\Delta G < 0$; $|T\Delta S| < |\Delta H|$
    (C) $\Delta G > 0$; $|T\Delta S| > |\Delta H|$
    (D) $\Delta G > 0$; $|T\Delta S| < |\Delta H|$
    (E) $\Delta G = 0$; $|T\Delta S| = |\Delta H|$

28. A 0.420 g sample of hemimellitic acid, $C_9H_6O_6$ (molar mass: 210 g), is neutralized by 0.020 L of 0.300 M NaOH. What is the number of protons per mole of $C_9H_6O_6$ available for donation to OH^- in water solution?

    (A) one
    (B) three
    (C) four
    (D) six
    (E) nine

29. What mass of water must be added to 100. g CH_3OH in order to make a solution that is 25% by mass CH_3OH?

    (A)   25.0 grams
    (B)   75.0 grams
    (C) 150. grams
    (D) 300. grams
    (E) 400. grams

30. Which of the following compounds has the lowest vapor pressure?

    (A) C_8H_{18}
    (B) $CHCl_3$
    (C) CH_3Cl
    (D) C_2H_5OH
    (E) $C_3H_5(OH)_3$

31. The vapor pressure of water is 80°C at 355 mmHg. Which expression gives the fraction of water molecules in a sample of nitrogen gas saturated with water vapor at 80°C and 740 mmHg total pressure?

    (A) $\dfrac{355}{740}$

    (B) $\dfrac{355}{760}$

    (C) $\dfrac{355}{355 + 740}$

    (D) $\dfrac{355}{355 + 760}$

    (E) $\dfrac{355 \times 18}{740 \times 28}$

32. Which structure best accounts for the properties of metals in the solid state?

    (A) atoms in fixed geometric positions
    (B) cations and anions in fixed geometric positions
    (C) multi-atom molecules in fixed geometric positions
    (D) cations in fixed geometric positions surrounded by a diffuse electron cloud
    (E) anions in fixed geometric positions surrounded by a diffuse electron cloud

33. According to the Kinetic Molecular Theory, which characteristic applies to an ideal gas?

    (A) An ideal gas has no critical temperature.
    (B) An ideal gas liquifies at temperatures below its critical temperature.
    (C) An ideal gas liquifies at temperatures above its critical temperature.
    (D) The critical temperature of an ideal gas is less than its triple point temperature.
    (E) The critical temperature of an ideal gas is greater than its triple point temperature.

34. Resonance helps to account for all of the following properties EXCEPT

   (A) the equal S$-$O bond energies in SO_2

   (B) the low reactivity of C in benzene, C_6H_6

   (C) the charge of 3^+ on the aluminum ion, Al^{3+}

   (D) the equal bond strengths in the nitrate ion, NO_3^-

   (E) the equal bond lengths in the carbonate ion, CO_3^{2-}

35. What is the shape of the SF_6 molecule?

   (A) linear
   (B) see-saw
   (C) octahedral
   (D) tetrahedral
   (E) trigonal bipyramidal

36. Which is formed when the nuclide $^{253}_{99}Es$ captures an alpha particle and emits a free neutron?

   (A) $^{258}_{98}Cf$

   (B) $^{256}_{99}Es$

   (C) $^{257}_{100}Fm$

   (D) $^{256}_{101}Md$

   (E) $^{258}_{101}Md$

37.

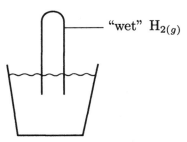

"wet" $H_{2(g)}$

A 50 mL sample of hydrogen gas is collected over water at 25°C and 745 mmHg as shown in the diagram above. All of the following describe properties of the system EXCEPT

   (A) The volume of $H_2O_{(g)}$ is equal to the volume of $H_{2(g)}$.

   (B) The pressure of $H_2O_{(g)}$ is equal to the pressure of $H_{2(g)}$.

   (C) The temperature of $H_2O_{(g)}$ is equal to the temperature of $H_{2(g)}$.

   (D) The number of molecules of $H_2O_{(g)}$ is less than the number of molecules of $H_{2(g)}$.

   (E) The average molecular velocity of $H_2O_{(g)}$ is less than the average molecular velocity of $H_{2(g)}$.

38. Which equation best illustrates the behavior of Fe^{3+} as a Lewis acid?

   (A) $Fe^{3+}_{(aq)} + SCN^-_{(aq)} \rightarrow FeSCN^{2+}_{(aq)}$

   (B) $Fe_2O_{3(s)} + FeO_{(s)} \rightarrow Fe_3O_{4(s)}$

   (C) $Fe^{3+}_{(aq)} + PO_4^{3-}_{(aq)} \rightarrow FePO_{4(s)}$

   (D) $Fe(NO_3)_{3(s)} \rightarrow Fe^{3+}_{(aq)} + 3NO_3^-_{(aq)}$

   (E) $Fe_{(s)} + \frac{3}{2}Cl_{2(g)} \rightarrow Fe^{3+}_{(aq)} + 3Cl^-_{(aq)}$

39. Equal masses of barium chloride dihydrate ($BaCl_2 \cdot 2H_2O$; molar mass: 244 g) and copper(II) sulfate pentahydrate ($CuSO_4 \cdot 5H_2O$; molar mass: 250 g) are placed in separate crucibles and heated until no further loss of mass is observed. Which expression is closest to the ratio of the mass of the contents of the $BaCl_2$ crucible to that of the $CuSO_4$ crucible after the heating is completed?

   (A) $\dfrac{36}{90}$

   (B) $\dfrac{160}{208}$

   (C) $\dfrac{208}{160}$

   (D) $\dfrac{244}{250}$

   (E) $\dfrac{208 + 160}{244 + 250}$

40. Which equation best illustrates the ionization behavior of liquid ammonia?

   (A) $NH_3 \rightleftharpoons N^{3-} + 3H^+$

   (B) $NH_3 \rightleftharpoons H^+ + NH_2^-$

   (C) $NH_3 + H_2O \rightleftharpoons NH_4^+ + OH^-$

   (D) $NH_3 + NH_3 \rightleftharpoons NH_4^+ + NH_2^-$

   (E) $NH_3 + H_2O \rightleftharpoons H_3O^+ + NH_2^-$

41. The normal boiling points of CH_3OH and H_2O are 65°C and 100°C, respectively. Which is another correct comparison of the properties of CH_3OH and H_2O?

     I.   The average molecular velocity of CH_3OH at 65°C is the same as the average molecular velocity of H_2O at 100°C.

     II.   The vapor pressure of CH_3OH at 65°C is the same as the vapor pressure of H_2O at 100°C.

     III.   The heat of vaporization of CH_3OH at 65°C is the same as the heat of vaporization of H_2O at 100°C.

(A) I only

(B) II only

(C) I and III only

(D) II and III only

(E) I, II, and III

42. Of the following pairs of substances, which should be chosen to react with $Al(OH)_3$ to demonstrate its behavior as an amphiprotic compound?

(A) oxygen and hydrogen

(B) ethanol and methanol

(C) sodium nitrate and sodium phosphate

(D) sodium hydroxide and hydrochloric acid

(E) potassium chloride and sodium chloride

43. Which describes the changes in the concentrations of all the solute ions present after 100 mL of 0.20 M $KI_{(aq)}$ is mixed with 100 mL of 0.10 M $BaI_{2(aq)}$?

	$[Ba^{2+}]$	$[K^+]$	$[I^-]$
(A)	remains the same	remains the same	decreases
(B)	decreases	decreases	remains the same
(C)	decreases	remains the same	remains the same
(D)	remains the same	decreases	decreases
(E)	decreases	remains the same	remains the same

44. Which species that contains S in the +6 oxidation state is most likely to be found in greatest concentration in a water solution that has pOH = 1.0?

(A) H_2SO_4

(B) $H_2S_2O_7$

(C) SO_3

(D) HSO_4^-

(E) SO_4^{2-}

45. During the electrolysis of a solution of potassium iodide, which process occurs at the anode?

   (A) oxidation of I^- to form I_2

   (B) oxidation of H_2O to form O_2

   (C) reduction of K^+ to form KOH

   (D) reduction of K^+ to form K

   (E) reduction of H_2O to form H_2

46. What is the formula of the oxide of manganese that contains the smallest percent oxygen by mass?

   (A) MnO

   (B) MnO_2

   (C) MnO_3

   (D) Mn_2O_3

   (E) Mn_3O_4

47. Which reaction involves neither oxidation nor reduction?

   (A) burning of tin in chlorine gas
   (B) ignition of iron in powdered sulfur
   (C) decomposition of potassium chlorate
   (D) dissolving of calcium carbonate in hydrochloric acid
   (E) catalytic decomposition of hydrogen peroxide on manganese dioxide

48.
$$2Al_{(s)} + 3Cl_{2(g)} \rightarrow 2AlCl_{3(s)}$$

   A mixture containing 0.60 mol Al in contact with 0.75 mol Cl_2 is ignited in a closed container. Assuming complete reaction, which gives the maximum quantity of $AlCl_3$ produced and the corresponding quantity of unconsumed reactant?

   (A) 0.45 mol $AlCl_3$ and 0.15 mol Al unreacted

   (B) 0.45 mol $AlCl_3$ and 0.15 mol Cl_2 unreacted

   (C) 0.50 mol $AlCl_3$ and 0.10 mol Al unreacted

   (D) 0.45 mol $AlCl_3$ and 0.10 mol Cl_2 unreacted

   (E) 0.50 mol $AlCl_3$ and 0.15 mol Cl_2 unreacted

49. Which describes some of the contents of the solution that results when 667 mL of 0.10 M $NaNO_3$ solution is mixed with 333 mL of 0.10 M $Ba(NO_3)_2$ solution?

	$[Na^+]$	$[NO_3^-]$
(A)	0.10 M	0.20 M
(B)	0.10	0.30
(C)	0.067	0.13
(D)	0.067	0.20
(E)	0.15	0.10

Questions 50-52: The substance AB is 20% dissociated in dilute water solution according to the equation

$$AB_{(aq)} \rightleftharpoons A^+_{(aq)} + B^-_{(aq)}$$

50. Which gives the predicted freezing point in °C for a 0.050 m solution of AB?

    (A) -1.86×0.040

    (B) -1.86×0.060

    (C) -1.86×0.070

    (D) -1.86×0.20

    (E) $-1.86 \times 0.20 \times 0.050$

51. Which gives the ratio of ions to molecules in this solution?

    (A) $\frac{1}{5}$

    (B) $\frac{1}{4}$

    (C) $\frac{1}{2}$

    (D) $\frac{2}{5}$

    (E) $\frac{2}{3}$

52. Which expression gives the value of K_i for this ionization in a 0.050 M solution?

    (A) $\dfrac{0.010 \times 0.010}{0.050}$

    (B) $\dfrac{0.010 \times 0.010}{0.040}$

    (C) $\dfrac{0.010 \times 0.010}{0.030}$

    (D) $\dfrac{0.010 \times 0.010}{0.20 \times 0.050}$

    (E) $\dfrac{0.20 \times 0.010 \times 0.010}{0.050}$

---

53. $$...Cr_2O_7^{2-} + ...C_2H_5OH + ...H^+ \rightarrow ...Cr^{3+} + ...CO_2 + ...H_2O$$

When the skeleton equation above is balanced using lowest integer coefficients, what is the sum of the coefficients?

    (A) 12

    (B) 18

    (C) 24

    (D) 30

    (E) 36

54. Which applies to a saturated solution of $Ca_3(PO_4)_{2(s)}$ in contact with excess $Ca_3(PO_4)_{2(s)}$?

$$Ca_3(PO_4)_{2(s)} \rightleftharpoons 3Ca^{2+}{}_{(aq)} + 2PO_4{}^{3-}{}_{(aq)} \qquad K_{sp} = 1.3 \times 10^{-32}$$

    I.   The free energy, G, of the products is equal to the free energy of the reactants.

    II. The rate of precipitation is equal to the rate of dissolving.

    III. The concentration of the reactants is equal to the concentration of the products.

(A) I only
(B) II only
(C) I and II only
(D) II and III only
(E) I, II, and III

55.

Acid	K_a
H_3PO_4	8×10^{-3}
$H_2PO_4{}^-$	8×10^{-8}
$HPO_4{}^{2-}$	5×10^{-13}

A mixture is prepared by adding 1.0 mole each of H_3PO_4, Na_2HPO_4 and NaH_2PO_4 to enough water to make 1.0 liter of solution. The $[H^+]$ in this solution is closest to

(A) 8×10^{-3}
(B) 5×10^{-4}
(C) 4×10^{-4}
(D) 2×10^{-8}
(E) 1×10^{-8}

56. The pH of 0.10 M solution of $NH_4C_2H_3O_2$ is closest to

(A) 1
(B) 3
(C) 7
(D) 11
(E) 14

57. $Ag_2CrO_{4(s)} \rightleftharpoons 2Ag^+_{(aq)} + CrO_4^{2-}_{(aq)}$

The molar solubility of $Ag_2CrO_{4(s)}$ is 1.3×10^{-4} mol L^{-1}. Which expression gives the value for K_{sp}, the solubility product constant?

(A) $(1.3 \times 10^{-4})(1.3 \times 10^{-4})^2$

(B) $(2.6 \times 10^{-4})^2(1.3 \times 10^{-4})$

(C) $(2.6 \times 10^{-4})(1.3 \times 10^{-4})$

(D $\dfrac{(2.6 \times 10^{-4})^2}{1.3 \times 10^{-4}}$

(E) $\dfrac{1.3 \times 10^{-4}}{(2.6 \times 10^{-4})^2}$

58. Which expression gives the percent H_3PO_4 (molar mass: 98 g) by mass in 2.0 molal solution of H_3PO_4 in water?

(A) $\dfrac{98 \times 100}{2 + 1000}$

(B) $\dfrac{98 \times 100}{98 + 1000}$

(C) $\dfrac{2 \times 98 \times 100}{98 + 1000}$

(D) $\dfrac{2 \times 98 \times 100}{1000}$

(E) $\dfrac{2 \times 98 \times 100}{(2 \times 98) + 1000}$

59. $NaI_{(s)} \rightleftharpoons Na^+_{(aq)} + I^-_{(aq)}$ $\Delta H = -7.5$ kJ

The equation above represents the solubility equilibrium in a saturated solution of NaI in contact with solid NaI. Which of the following is the best way to increase the rate at which the dissolving process occurs?

(A) adding $NaI_{(s)}$

(B) adding saturated $NaI_{(aq)}$

(C) cooling the mixture

(D) warming the mixture

(E) increasing the external pressure

60. Which species should be used to precipitate Pb^{2+}, as the only cation, from a mixture of Pb^{2+} and Ca^{2+} ions in aqueous solution?

(A) Cl^-

(B) NO_3^-

(C) NH_4^+

(D) H_3O^+

(E) CH_3COO^-

61.
$$CH_3OH_{(\ell)} + 2O_{2(g)} \rightarrow CO_{2(g)} + 2H_2O_{(\ell)} \qquad \Delta H^\circ = ?$$

$$\Delta H^\circ_f \text{ for } H_2O_{(\ell)} = -300 \text{ kJ mol}^{-1}$$

$$\Delta H^\circ_f \text{ for } CO_{2(g)} = -400 \text{ kJ mol}^{-1}$$

$$\Delta H^\circ_f \text{ for } CH_3OH_{(\ell)} = -300 \text{ kJ mol}^{-1}$$

The approximate standard heats of formation for water, carbon dioxide, and methanol, are given above. Based on this information, the standard heat of combustion for methanol in kJ mol^{-1} is closest to

(A) +1,700

(B) zero

(C) −300

(D) −700

(E) −1,700

62. Which of the following has the least effect on the rate of a reaction?

(A) adding a solid catalyst to a gas phase reaction

(B) adding a solid catalyst to a liquid phase reaction

(C) adding inert miscible liquid to a liquid phase reaction

(D) adding inert gas to a gas phase reaction at constant volume

(E) adding excess base to neutralization reaction for a nearly insoluble weak acid

63.   ethene $(C_2H_{4(g)})$ $\Delta G^\circ_f = 68 \text{ kJ mol}^{-1}$ $\Delta H^\circ_f = 52 \text{ kJ mol}^{-1}$

ethyne $(C_2H_{2(g)})$ $\Delta G^\circ_f = 209 \text{ kJ mol}^{-1}$ $\Delta H^\circ_f = 227 \text{ kJ mol}^{-1}$

Based on the information above, what can be concluded about the relative stability and the standard entropies of formation of these compounds at 298 K?

(A) Both compounds have the same stability and the same sign for ΔS°_f.

(B) Ethene is more stable than ethyne and both have the same sign for ΔS°_f.

(C) Ethene is less stable than ethyne and both have the same sign for ΔS°_f.

(D) Ethene is more stable than ethyne and they have opposite signs for ΔS°_f.

(E) Ethene is less stable than ethyne and they have opposite signs for ΔS°_f.

64. Information about NaCl is given below

$NaCl_{(s)}$ $\Delta H_f^\circ = -410$ kJ mol^{-1} $\Delta G_f^\circ = -384$ kJ mol^{-1}

$NaCl_{(aq)}$ $\Delta H_f^\circ = -407$ kJ mol^{-1} $\Delta G_f^\circ = -393$ kJ mol^{-1}

Which range includes the value in J mol^{-1} K^{-1} for ΔS° for the dissolving process at 298 K?

(A) greater than 30 J mol^{-1} K^{-1}

(B) 1 – 30 J mol^{-1} K^{-1}

(C) 0.1 – 1 J mol^{-1} K^{-1}

(D) 0.001 – 0.1 J mol^{-1} K^{-1}

(E) less than 0.001 J mol^{-1} K^{-1}

Questions 65 and 66: $2A + B + C \rightarrow$ products

Four trials of the reaction above were carried out in order to determine its rate law. The following data were collected.

Trial	[A]	[B]	[C]	Initial rate M sec^{-1}
1	0.02	0.02	0.02	1.6×10^{-3}
2	0.01	0.02	0.02	8.0×10^{-4}
3	0.01	0.04	0.02	1.6×10^{-3}
4	0.01	0.04	0.03	1.6×10^{-3}

65. Based on these observations, what is the rate law?

(A) Rate = k[A]^2

(B) Rate = k[B][C]

(C) Rate = k[A][B]

(D) Rate = k[A]^2[B]^2

(E) Rate = k[A]^2[B][C]

66. As any trial of this reaction proceeds at constant temperature, the rate of the reaction

(A) remains the same because no catalyst is added

(B) remains the same because the temperature is constant

(C) increases because the rate constant is a positive number

(D) decreases because the concentrations of the reactants decrease

(E) decreases because the effectiveness of collisions between molecules decreases

67. Which of the following identifies a pair of isomers of diethyl ether?

    (A) 1-propanol and 1-pentanol
    (B) 2-butanol and 2-methyl-2-propanol
    (C) 2-butanol and 2-methyl-1-butanol
    (D) 2-methyl-1-butanol and 2-methyl-2-butanol
    (E) 2-methyl-1-propanol and 2-methyl-1-pentanol

68. Which is the best net ionic equation for the reaction of a piece of mossy zinc with excess dilute sulfuric acid?

    (A) $Zn_{(s)} + H_2SO_{4(aq)} \rightarrow H_{2(g)} + ZnSO_{4(s)}$

    (B) $Zn^{2+}{}_{(aq)} + HSO_4^-{}_{(aq)} \rightarrow Zn(HSO_4)_{2(s)}$

    (C) $Zn^{2+}{}_{(aq)} + SO_4^{2-}{}_{(aq)} \rightarrow ZnSO_{4(s)}$

    (D) $Zn_{(s)} + 2\ H^-{}_{(aq)} \rightarrow ZnH_{2(s)}$

    (E) $Zn_{(s)} + 2\ H_3O^+{}_{(aq)} \rightarrow Zn^{2+}{}_{(aq)} + H_{2(g)} + 2\ H_2O_{(\ell)}$

69. In the correct IUPAC name for the molecule,

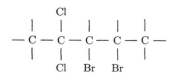

    the prefix, di, is used .?. and the numeral, 4, is used .?.

    (A) never, twice
    (B) once, once
    (C) once, twice
    (D) twice, once
    (E) twice, twice

70. What molecular formula is represented by the structural formula

    (A) $C_4N_2O_4$
    (B) $C_6N_2O_4$
    (C) $C_4H_4N_2O_4$
    (D) $C_6H_4N_2O_4$
    (E) $C_6H_6N_2O_4$

71. The diagram below represents a sample of pure water in beaker A and an equal volume of a solution of sugar in water in beaker B. Both beakers are placed under a bell jar. Which of the following changes is most likely to occur as this apparatus is observed over several weeks?

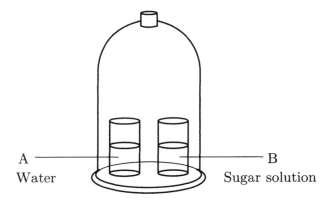

A ——————— Water

B ——————— Sugar solution

(A) The contents of both beakers evaporate to dryness.
(B) The volume of liquid in beaker B becomes greater than the volume of liquid in beaker A.
(C) The mass of the solute in beaker B decreases as the mass of the solute in beaker A increases.
(D) The concentration of the solute in beaker B decreases as the mass of the solute in beaker A increases.
(E) The concentration of the solute in beaker B increases as the volume of liquid in beaker A increases.

72. In a mercury cell, often used as a calculator battery, which element is used as the anode?

(A) carbon
(B) chromium
(C) manganese
(D) sodium
(E) zinc

73. In the ordinary use of a styrofoam cup as a calorimeter for an exothermic dissolving reaction, which property requires the least precision and accuracy in its measurement?

(A) final temperature of the solution
(B) initial temperature of the solvent
(C) mass of water as solvent is added to a calorimeter
(D) mass of the solute that is added to the calorimeter
(E) change in volume of the system as the reaction proceeds.

74. When pure (18 M) sulfuric acid dissolves in water, a significant increase in the temperature of the system occurs. Which gives the correct signs for the thermodynamic parameters for this dissolving process?

	ΔG_{soln}	ΔH_{soln}	ΔS_{soln}
(A)	+	+	+
(B)	+	+	−
(C)	−	−	+
(D)	−	+	+
(E)	−	−	−

75. Which gives the thermodynamic parameters for the phase change in a system that begins as an open container of liquid water placed in a constant temperature environment of 230 K?

$$\text{I.} \quad \Delta G < 0$$
$$\text{II.} \quad \Delta H < 0$$
$$\text{III.} \quad \Delta S < 0$$

(A) I only
(B) III only
(C) I and III only
(D) II and III only
(E) I, II, and III

Section II

Section II - Free Response   Total Time – 90 Minutes
(Multiple-Choice Questions are found in Section I.)

Part A: Question 76
and
Question 77 or Question 78
Time: 40 minutes

Access to calculators, Periodic Table, lists of standard reduction potentials, and
Equations and Constants

(2004 Examination directions) Clearly show the method used and the steps involved in
arriving at your answers. It is to your advantage to do this, because you may obtain
partial credit if you do and you will receive little or no credit if you do not. Attention
should be paid to significant figures. Be sure to write all your answers to the questions
on the lined pages following each question in the booklet with the pink cover. Do not
write your answers on the green insert.

Answer question 76 below. The Section II score weighting for this question is 20 percent.

76. Answer all four questions below related to the dissolving of calcium hydroxide in
     water. The solubility of $Ca(OH)_2$ is 0.51 grams/liter at 298 K.

     (a) Write the balanced equation for the solubility equilibrium of $Ca(OH)_2$ in water
          and the corresponding mass action expression for the solubility product
          constant, K_{sp}.

     (b) Calculate the value for the solubility product constant, K_{sp}, at 298 K.

     (c) Calculate the pH of a saturated solution of $Ca(OH)_2$ at 298 K.

     (d) A mixture is prepared by adding 40.0 mL of 0.020 M $CaCl_{2(aq)}$ to 60.0 mL of
          0.015 M $KOH_{(aq)}$. Does precipitation of $Ca(OH)_{2(s)}$ occur at 298 K? Show
          calculations to support your answer. (Assume the volumes of the two solutions
          are additive.)

Answer either question 77 or question 78 below.

(2004 examination directions) Only one of these two questions will be graded. If you start both questions, be sure to cross out the question you do not want graded.

The Section II score weighting for the question that you choose is 20 percent.

77. Answer all four questions below about the combustion (burning) of butane. Butane is a hydrocarbon fuel gas used for many camp stoves and lanterns.

(a) Write the balanced equation for the complete combustion of butane, $C_4H_{10(g)}$, in oxygen at 298 K. Include the phase for each reactant and product in its standard state.

(b) What volume of $O_{2(g)}$, measured at 0.965 atm and 20.0°C, is consumed when 10.0 g $C_4H_{10(g)}$ is burned?

(c) Some standard heats of formation, ΔH°_f, at 298 K are given below

$$H_2O_{(\ell)} \qquad -285.83 \text{ kJ mol}^{-1}$$

$$C_4H_{10(g)} \qquad ..?.. \text{ kJ mol}^{-1}$$

$$CO_{2(g)} \qquad -393.5 \text{ kJ mol}^{-1}$$

The standard heat of combustion, ΔH°_{comb}, of butane at 298 K is $-2,874.5$ kJ mol^{-1}. Determine the standard heat of formation, ΔH°_f, of butane.

(d) Energy from the burning of butane can be used to heat water. Assuming all the energy produced is transferred to the water, what mass of water can be heated from 15.0°C to 70.0°C when 1.00 mole of $C_4H_{10(g)}$ is burned? (The specific heat capacity of water is 4.18 J g^{-1} °C^{-1}.)

78. Answer all four questions below about the chemical reaction for which rate information is given.

The reaction between $NO_{(g)} + Br_{2(g)}$ to form $NOBr_{(g)}$ is studied in the four trials described below:

$$2NO_{(g)} + Br_{2(g)} \rightarrow 2NOBr_{(g)}$$

Trial	Initial [NO] $(mol\ L^{-1})$	Initial [Br$_2$] $(mol\ L^{-1})$	Initial rate of formation of NOBr $(mol\ L^{-1}\ sec^{-1})$
I	0.100	0.100	1.20×10^{-3}
II	0.100	0.200	2.40×10^{-3}
III	0.200	0.100	4.80×10^{-3}
IV	0.300	0.100	1.08×10^{-2}

(a) Write the rate law for the reaction above in the form $Rate = k[NO]^x[Br_2]^y$. Explain how you determined the values for exponents x and y.

(b) On the basis of the rate law determined in part (A), calculate the specific rate constant. Specify the units.

(c) Calculate the initial rate in trial III if the volume of the container for the reaction mixture is decreased to half its original volume.

(d) Calculate the rate in trial IV after [NO] has decreased to 0.150 M.

Part B: Questions 79, 80, 81 and
Question 82 or Question 83
Time: 50 minutes

Access to Periodic Table, lists of standard reduction potentials
and *Equations and Constants*
No access to calculators

Answer question 79 below: The Section II score weighting for this question is 15 percent.

79. (2004 Examination directions) Write the formulas to show the reactants and the products for any FIVE of the laboratory situations described below. Answers to more than five choices will not be graded. In all cases, a reaction occurs. Assume that the solutions are aqueous unless otherwise indicated. Represent substances in solution as ions if the substances are extensively ionized. Omit formulas for any ions or molecules that are unchanged by the reaction. You need not balance the equations.

(a) Diethyl ether is ignited in air.

(b) A dilute solution of hydrofluoric acid is added to excess potassium hydroxide solution.

(c) Solutions of sodium sulfide and manganese(II) nitrate are mixed.

(d) Solid calcium oxide is sprinkled onto water.

(e) An electric current is passed through molten potassium iodide.

(f) Solid potassium chlorate is mixed with solid manganese(IV) oxide and heated in a test tube.

(g) Fine iron wire is heated in oxygen.

(h) A bar of aluminum is placed in a solution of copper(II) nitrate.

(2004 Examination directions) Your responses to the rest of the questions in this part of the examination will be graded on the basis of the accuracy and relevance of the information cited. Explanations should be clear and well organized. Examples and equations may be included in your responses where appropriate. Specific answers are preferable to broad, diffuse responses.

(2004 examination directions) **Answer both Question 80 and Question 81 below.**
Both questions will be graded.

The Section II score weighting for these questions is 30 percent (15 percent each).

80.

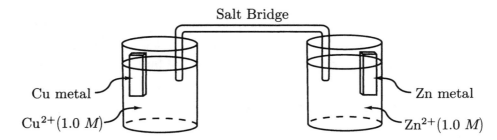

Answer the following questions about the setup and operation of a galvanic cell.

(a) One item is missing from this electrochemical cell that prevents it from functioning. Identify that item by drawing it onto the diagram. Label the missing item.

(b) Once this cell is operating, a redox reaction occurs.

   (i) Which electrode, Cu or Zn, is the cathode? Explain.

   (ii) Write the reaction for the half-cell containing zinc.

   (iii) Calculate the voltage E° for this cell.

(c) Set up the equation to calculate E in volts for the cell when the molarity of Cu^{2+} in the left half-cell is reduced to 0.10 M. Label each factor in the equation with proper units.

(d) Suppose that the temperature of the original galvanic cell is decreased to 273 K. Would the voltage decrease, stay the same, or increase? Explain.

(e) What would be the effect on the voltage if the metal electrodes were doubled in mass? Explain.

(f) What would be the effect on the voltage if a solution of NaOH were poured into the Zn/Zn^{2+} half-cell? Explain.

(g) What would be the effect on the voltage if a solution of 10.0 M $Zn(NO_3)_2$ were poured into the Zn/Zn^{2+} half-cell? Explain.

81. Answer the four questions below about a buffer solution

$$H_2O + CO_2 \rightleftharpoons HCO_3^- + H^+ \qquad K_I = 4.4 \times 10^{-7}$$
$$HCO_3^- \rightleftharpoons CO_3^{2-} + H^+ \qquad K_{II} = 4.7 \times 10^{-11}$$

A buffer solution is prepared by placing 0.30 mol CO_3^{2-} and 0.50 mol HCO_3^- in sufficient water to prepare 1.00 liter of solution. Half of this solution is poured into a flask labeled #1; the other half is placed in flask #2.

(a) What is a buffer solution? What characteristics of the dissolved species give this solution its buffering properties?

(b) Draw Lewis dot structures for hydrogen carbonate ion and hydroxide ion.

(c) What is the effect on $[CO_3^{2-}]$ and $[HCO_3^-]$ when 0.10 mol $KHSO_{4(s)}$ is added to the solution in flask #1? Account for your answer in terms of moles of protons transferred. Has the capacity of the buffer been exceeded?

(d) In another experiment 0.10 mol $NaOH_{(s)}$ is added to the solution in flask #2. What are the new equilibrium values for $[CO_3^{2-}]$ and $[HCO_3^-]$?

(2004 examination directions) **Answer either question 82 or question 83 below.** Only one of these two questions will be graded. If you start both questions, be sure to cross out the question you do not want graded. The Section II score weighting for the question that you choose is 15 percent.

82. Answer all four questions below about the setup and operation of an electrolytic device.

    The term, faraday, refers to the quantity of charge on one mole of electrons. That value can be determined experimentally by the electrolysis of a dilute solution of sulfuric acid and collecting the gases produced. The materials and apparatus available are listed below:

apparatus to be assembled	other apparatus
ammeter	barometer
dilute sulfuric acid in a large vessel	clock
platinum wires	thermometer
source of direct current	
two graduated measuring tubes	

    (a) Sketch a diagram to show how to assemble the apparatus listed above to conduct the experiment that will provide a value of the faraday. You may choose to connect the apparatus in the diagram below.

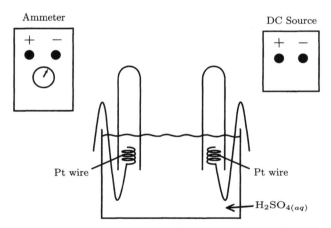

    Sulfuric acid solution with two gas measuring tubes and two platinum wires

    (b) When the volume of hydrogen collected is 90.0 mL, what is the expected volume of oxygen? Explain.

    (c) What measurements are necessary to determine the number of coulombs of electricity that are transferred during the experiment? Explain.

    (d) How would the experiment be affected if copper metal were used in place of platinum? Explain.

83. Provide the information specified in the three questions below about the classification of solids.

Most crystalline solids can be classified into one of the four categories specified in columns 1 through 4 listed below. This classification is based upon differences in force of attraction between the particles found at the lattice points.

Answer each of the following for the four categories of solid listed according to the directions below. For each category, one piece of information is given. Complete the table by writing your answer in each space provided.

   (a)   Identify a substance that corresponds to the given information in columns 2 and 4.

   (b)   For columns 1, 3 and 4, specify the chemical symbol or formula for the particle(s) found at each lattice point for the example given on line (a) in the table.

   (c)   For columns 1, 2 and 3, identify or describe the force of attraction between the particles specified on line (b) the table.

	Categories of Solids			
	1	2	3	4
	ionic solid	metallic solid	molecular solid	network (covalent solid)
(i) example	CaS (calcium sulfide)		CO_2	
(ii) particles at lattice point		Mg^{2+} ion		
(iii) force between particles				covalent bond (shared pair of electrons)

SAMPLE EXAMINATION IV

Section I – Multiple Choice

Questions 1-5. The set of lettered choices, a list of oxides, below refers to the numbered phrases immediately following it. Select the one lettered choice that is most closely associated with each numbered phrase. Each lettered choice can be used once, more than once or not at all.

(A) SO_2

(B) BaO_2

(C) CO_2

(D) GeO_2

(E) NO_2

1. an odd electron molecule

2. an ionic compound

3. at STP the gas that illustrates greatest deviation from ideal behavior

4. source of a semiconductor

5. includes a element with oxidation number of +2

---

6. Which event is most likely to occur in an experiment to measure ionization energy?
   (A) A positive ion is converted to a negative ion.
   (B) A neutral atom is converted to a positive ion.
   (C) A neutral atom is converted to a negative ion.
   (D) A negative ion is converted to a neutral atom.
   (E) A negative ion is converted to a positive ion.

7. All of these sets of quantum numbers apply to an electron in the *p*-sublevel EXCEPT

   (A) $2, \ 1, \ 1, \ +\frac{1}{2}$

   (B) $3, \ 1, \ 0, \ +\frac{1}{2}$

   (C) $3, \ 1, \ 0, \ -\frac{1}{2}$

   (D) $2, \ 0, \ 0, \ +\frac{1}{2}$

   (E) $2, \ 1, \ 0, \ +\frac{1}{2}$

8. Which is a correct comparison of a sulfide ion to a sulfur atom?

       I.   The radius of the sulfur atom is greater.

      II.   The sulfide ion contains more electrons.

     III.   The number of energy levels occupied by electrons is the same.

   (A) I only

   (B) II only

   (C) I and II only

   (D) I and III only

   (E) II and III only

Questions 9-11: The set of lettered choices below is a list of molecular formulas for certain gases. Select the one lettered choice that best fits each numbered description of the bonds within the molecules of the gas.

   (A) H_2

   (B) N_2

   (C) O_2

   (D) F_2

   (E) Cl_2

9. contains bond with greatest multiplicity

10. has the strongest bond

11. has the shortest bond length

Questions 12-14: The set of lettered choices below is a list of molecular geometries. For each numbered species, select the one lettered choice that describes its molecular geometry.

(A) linear

(B) seesaw

(C) square planar

(D) square pyramidal

(E) T-shaped

12. $XeCl_4$

13. I_3^-

14. IF_3

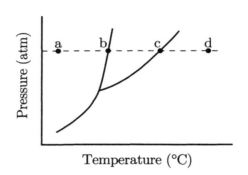

Temperature (°C)

15. The dashed line, ad, in the phase diagram above represents properties of a closed system as energy is added to that system at a constant rate. The properties are observed at points a, b, c, and d. Which is associated with the longest time period?

(A) change in temperature from a to b

(B) equilibrium with constant temperature at b

(C) change in temperature from b to c

(D) equilibrium with constant temperature at c

(E) change in temperature from c to d

16. Which gives a reason why, compared to CO_2, SO_2 exhibits greater deviation from ideal gas behavior?

      I.   The O-S-O bond angle is greater than the O-C-O bond angle.

     II.   A molecule of SO_2 contains more electrons than a molecule of CO_2.

    III.   The bond order of the S-O bond is less than the bond order of the C-O bond.

(A) I only

(B) II only

(C) III only

(D) I and III only

(E) II and III only

17. When these five water solution systems are listed in order of increasing vapor pressure, which position is occupied by $0.1\ m\ C_6H_{12}O_6$, glucose? (Assume ideal behavior.)

    $0.2\ m\ KNO_3$

    $0.2\ m\ (NH_2)_2CO$, urea

    $0.1\ m\ CaCl_2$

    $0.1\ m\ C_2H_5OH$

    $0.1\ m\ C_6H_{12}O_6$, glucose

(A) first

(B) second

(C) third

(D) fourth

(E) fifth

18. A solution is prepared by dissolving 8.01 g of ammonium nitrate, NH_4NO_3 (molar mass: 80.1 g), in enough water to yield 250.0 mL of solution. What is the molarity of ammonium nitrate in this solution?

(A) $32.0\ M$

(B) $4.00\ M$

(C) $0.400\ M$

(D) $0.100\ M$

(E) $0.0400\ M$

19. Which aqueous solution has the highest boiling point?

    (A) 0.1 m $SrBr_2$

    (B) 0.1 m KBr

    (C) 0.1 m $MgSO_4$

    (D) 0.1 m CH_3COOH

    (E) 0.1 m C_2H_5OH

20. Which of these chlorine-containing compounds is most soluble in water?

    (A) AgCl

    (B) CCl_3OH

    (C) CCl_4

    (D) $PbCl_2$

    (E) $HClO_4$

21. Which conditions of pressure and temperature favor greatest solubility of a gas into a liquid?

	P	T
(A)	low	low
(B)	low	high
(C)	high	low
(D)	high	high
(E)	moderate	moderate

22. A solution is prepared by dissolving 1.0 mol of acetic acid, CH_3COOH (molar mass: 60.0 g), in 5.0 mol water (molar mass: 18 g). Which expression gives the molality of this solution.

    (A) $\dfrac{1}{0.090}$

    (B) $\dfrac{1}{5}$

    (C) $\dfrac{1}{6}$

    (D) $\dfrac{60}{5 \times 18}$

    (E) $\dfrac{60}{0.090}$

23. Consider a solution that is 0.50 X (mole fraction) of ethanol (molar mass: 46 g) in water (molar mass: 18 g). Which value gives the best approximation for percent by mass ethanol in that solution?

    (A) 10%

    (B) 25%

    (C) 50%

    (D) 75%

    (E) 90%

24. The mass percent of one oxide of manganese is determined to be 69.6% Mn and 30.4% O. Which expression is the best representation of the empirical formula of this compound?

    (A) $Mn_{\frac{69.6}{6.02}}$ $O_{\frac{30.4}{6.02}}$

    (B) $Mn_{\frac{69.6}{30.4}}$ $O_{\frac{30.4}{30.4}}$

    (C) $Mn_{\frac{69.6}{30.4}}$ $O_{\frac{30.4}{69.6}}$

    (D) $Mn_{\frac{69.6}{55.0}}$ $O_{\frac{30.4}{16.0}}$

    (E) $Mn_{\frac{55.0}{69.6}}$ $O_{\frac{16.0}{30.4}}$

25. When zinc reacts with nitric acid, one product is ammonium nitrate. When the corresponding half-reaction as shown below is balanced using the lowest integers, what is the sum of the coefficients?

    $$..?..\ NO_3^- \ +\ ..?..\ H^+ \ +\ ..?..\ e^- \ \rightarrow\ ..?..\ NH_4^+ \ +\ ..?..\ H_2O$$

    (A) 13

    (B) 15

    (C) 19

    (D) 21

    (E) 23

26. Consider a mixture of gases that contains 0.10 mol of $N_2O_{5(g)}$ and 0.10 mol of $NO_{2(g)}$ at STP. Which gives a correct description of a quantity of material present?

        I.   The number of atoms is greater than 5×10^{23}.

        II.   The number of molecules is greater than 1×10^{23}.

        III.   The volume of the sample is greater than 2.24 liters.

    (A) I and II only

    (B) II and III only

    (C) III only

    (D) I and III only

    (E) I, II and III only

27. Consider the reaction of powdered zinc metal with hydrochloric acid to form hydrogen gas. Each of the following causes the reaction rate to decrease EXCEPT

   (A) substituting a zinc bar of the same mass as the powdered zinc

   (B) cooling the reaction mixture

   (C) diluting the acid solution

   (D) stirring the reaction mixture

   (E) substituting HF solution of equal molarity for the HCl solution

28. Ordinarily the rate of any reaction in a closed system at constant temperature decreases over time. Which corresponding change over time accounts for this phenomenon?

   (A) The concentrations of the products increase.

   (B) The concentrations of the reactants decrease.

   (C) The number of effective collisions between reacting particles increases.

   (D) The energy of activation decreases.

   (E) The effective concentration of the catalyst decreases.

29. Strontium-90 has a half-life of 28.8 years. Approximately what mass from a 50.0 gram sample of pure ^{90}Sr would remain after the passage of 90 years?

   (A) 33.3 g

   (B) 25.0 g

   (C) 16.7 g

   (D) 12.5 g

   (E) 6.25 g

30. Carbon-14 has a half-life of approximately 6000 years. What fraction of an original sample would remain after 24,000 years?

   (A) 1/64

   (B) 1/32

   (C) 1/16

   (D) 1/8

   (E) 1/4

31. Each of the following is a reasonable label for the rate of a reaction EXCEPT

   (A) $mol\ L^{-1}$

   (B) $mol\ L^{-1}\ sec^{-1}$

   (C) $molecules\ sec^{-1}$

   (D) $g\ sec^{-1}$

   (E) $mol\ L^{-1}\ sec^{-1}$

32. When the rate law for a reaction is second order in terms of a certain reactant, how will the reaction rate be affected when the concentration of that reactant is doubled?

   (A) halved

   (B) remains the same

   (C) doubles

   (D) triples

   (E) quadruples

33. Excess 0.1 M $NaCl_{(aq)}$ is added to 100 mL of 0.1 M $AgNO_{3(aq)}$. A white precipitate forms. Which is a correct description of the quantities present?

	liquid phase	**solid phase**
(A)	$[Ag^+] = [Cl^-]$	mol Ag^+ = mol Cl^-
(B)	$[Ag^+] < [Cl^-]$	mol Ag^+ = mol Cl^-
(C)	$[Ag^+] = [Cl^-]$	mol Ag^+ < mol Cl^-
(D)	$[Ag^+] < [Cl^-]$	mol Ag^+ < mol Cl^-
(E)	$[Ag^+] > [Cl^-]$	mol Ag^+ < mol Cl^-

34. Which statement of equality applies to the equilibrium established when excess 0.1 M $NaCl_{(aq)}$ is added to 100 mL of 0.1 M $AgNO_{3(aq)}$?

   (A) $[Na^+] = [NO_3^-]$

   (B) $[Ag^+] = [Cl^-]$

   (C) $[Na^+] = [Cl^-]$

   (D) The rate of formation of $AgCl_{(s)}$ = rate of dissolution of $AgCl_{(s)}$

   (E) The rate of formation of $NaCl_{(s)}$ = rate of formation of $AgCl_{(s)}$

35. A sealed metal tank contains the equilibrium system represented by the equation below.

$$2NO_{(g)} + \tfrac{1}{2}O_{2(g)} \rightleftharpoons N_2O_{3(g)} + 90 \text{ kJ}$$

Some additional oxygen is added to the equilibrium system and the tank resealed. A new equilibrium is achieved after 10 minutes. Conditions of pressure and temperature are represented by the symbols in the table below.

	pressure	temperature
at the original equilibrium	P_1	T_1
at the time oxygen added	P_2	T_2
at the final equilibrium	P_3	T_3

Which is a correct statement of the relationship between these values?

(A) P_1 is the greatest of the three pressures; T_1 is the greatest of the three temperatures.

(B) P_3 is the greatest of the three pressures; T_1 is the greatest of the three temperatures.

(C) P_2 is the greatest of the three pressures; T_1 is the greatest of the three temperatures.

(D) P_2 is the greatest of the three pressures; T_3 is the greatest of the three temperatures.

(E) P_3 is the greatest of the three pressures; T_3 is the greatest of the three temperatures.

36. For which ranges of values for enthalpy change, ΔH, and entropy change, ΔS, is a reaction always spontaneous?

	ΔH	ΔS
(A)	>0	>0
(B)	>0	<0
(C)	<0	>0
(D)	0	<0
(E)	>0	0

37.
$$C_5H_{12(\ell)} + 8O_{2(g)} \rightarrow 5CO_{2(g)} + 6H_2O_{(g)}$$

For the combustion of pentane as shown in the equation above, what are the signs of ΔG, ΔH, and ΔS?

	ΔG	ΔH	ΔS
(A)	−	−	−
(B)	−	−	+
(C)	−	+	−
(D)	+	−	+
(E)	+	+	+

38. Which terms describe the quantity of energy in the universe and the measure of entropy in the universe, respectively?

   (A) constant, increasing

   (B) constant, constant

   (C) constant, decreasing

   (D) increasing, constant

   (E) decreasing, decreasing

39. Which applies to a reaction system for which $\Delta G = 0$?

   (A) The reaction system has gone to completion.

   (B) The reaction system has reached equilibrium.

   (C) The reaction system has reached the temperature of 0 kelvins.

   (D) The entropy of the reaction system has reached zero.

   (E) The limiting reactant has been consumed.

40. Which properties must be known in order to use a constant-pressure ("coffee cup") calorimeter to investigate heats of reaction?

   I.   Heat capacity of the calorimeter, c_{cal}

   II.   Heat capacity of the solution, $c_{sol'n}$

   III. Mass of solution

   (A) I only

   (B) II only

   (C) III only

   (D) I and III only

   (E) I, II, and III

41. Which 0.1 M solution has the lowest pH?

   (A) $NaNO_3$

   (B) NaH_2PO_4

   (C) $NaOH$

   (D) Na_2CO_3

   (E) $NaHCO_3$

42. Which pair of equations illustrates the amphiprotic properties of some aluminum-containing species?

   1. $Al^{3+}_{(aq)} + S^{2-}_{(aq)} \rightarrow Al_2S_{3(s)}$

   2. $4Al_{(s)} + 3O_{2(g)} \rightarrow 2Al_2O_{3(s)}$

   3. $2Al_{(s)} + Fe_2O_{3(s)} \rightarrow Al_2O_{3(s)} + 2Fe_{(s)}$

   4. $AlCl_{3(s)} + NaCl_{(aq)} \rightarrow AlCl_4^-_{(aq)} + Na^+_{(aq)}$

   5. $Al(OH)_{3(s)} + 3H^+_{(aq)} \rightarrow Al^{3+}_{(aq)} + 3H_2O$

   6. $Al(OH)_{3(s)} + OH^-_{(aq)} \rightarrow AlO_2^-_{(aq)} + 2H_2O$

   (A) equations 1 & 4
   (B) equations 2 & 3
   (C) equations 2 & 6
   (D) equations 4 & 5
   (E) equations 5 & 6

43. Which accounts for the chemical change that occurs when $NH_{3(aq)}$ is added to $HCl_{(aq)}$?

   (A) The Cl^- ion gives up excess electrons.

   (B) When dissolved in water, the Cl^- ion takes on eight valence electrons.

   (C) The NH_3 molecule has a greater dipole moment than the HCl molecule.

   (D) The H-N-H bond angle is less than 109°47' (the regular tetrahedral bond angle).

   (E) The unshared electron pair on NH_3 provides a bond site for the transferred proton.

44. Which applies to the process of the electrolysis of a dilute H_2SO_4 solution?

   I. The pH of the solution around the anode becomes lower.
   II. Reduction occurs at the cathode.
   III. Bubbles appear at each electrode.

   (A) I only
   (B) II only
   (C) III only
   (D) I and III only
   (E) I, II, and III

45. Which description applies to any galvanic (voltaic) cell?

    (A) Oxidation occurs at the anode.

    (B) Electrons move from the cathode to the anode through an external circuit.

    (C) Electrons move from the anode to the cathode through the salt bridge.

    (D) The anode gains mass.

    (E) The cathode attracts anions.

46. Which describes the behavior of the galvanic (voltaic) cell
    $Zn/Zn^{2+}(1\ M)//Cu^{2+}(1\ M)/Cu$ during the discharge reaction?

    (A) Oxidation occurs at the copper electrode.

    (B) The mass of the zinc electrode decreases.

    (C) The concentration of Cu^{2+} ion increases.

    (D) Bubbling of gas occurs at each electrode.

    (E) The concentration of Zn^{2+} ion decreases.

47. For which reduction does the transfer of one mole of electrons produce the greatest mass of metal?

    (A) aluminum: $Al^{3+} \rightarrow Al$

    (B) chromium: $Cr^{3+} \rightarrow Cr$

    (C) copper: $Cu^{2+} \rightarrow Cu$

    (D) silver: $Ag^{+} \rightarrow Ag$

    (E) zinc: $Zn^{2+} \rightarrow Zn$

48. Which mathematical expression identifies the number of coulombs necessary to produce one mole of iron metal from an aqueous solution of iron(III) sulfate, $Fe_2(SO_4)_3$?

    (A) $\dfrac{96,500}{(2 \times 3)}$

    (B) $\dfrac{96,500}{3}$

    (C) $96,500$

    (D) $96,500 \times 3$

    (E) $96,500 \times (2 \times 3)$

49. Which describes a dead battery?

   I. $\Delta G = 0$

   II.   All chemical reactions have stopped.

   III. The cell can no longer do work.

(A) I only

(B) II only

(C) I and III only

(D) II and III only

(E) I, II, and III

50. Which gives the number of isomers by category for the hydrocarbon pentene, C_5H_{10}?

	straight chain isomers	branched chain isomers
(A)	1	2
(B)	1	3
(C)	2	2
(D)	2	3
(E)	2	4

51. Which set of hybrid orbitals accounts for the O–C–C bond angle in propanone, CH_3COCH_3?

(A) sp^2

(B) sp^3

(C) sp^4

(D) dsp^3

(E) d^2sp^3

52. Which six-carbon compound reacts most readily with $Br_{2(\ell)}$?

(A) glucose, $C_6H_{12}O_6$

(B) benzene, C_6H_6

(C) cyclohexene, C_6H_{10}

(D) cyclohexane, C_6H_{12}

(E) hexane, C_6H_{14}

Questions 53-56: The list of lettered choices below gives the symbols for five of the noble gases. For each numbered phrase, choose the symbol of the noble gas with which that phrase is most closely associated. A choice may be used once, more than once or not at all.

(A) He

(B) Ne

(C) Ar

(D) Xe

(E) Rn

53. first discovered in spectroscopy of the Sun

54. most abundant noble gas in the atmosphere

55. name means "the stranger"

56. prolonged exposure causes lung cancer

---

57. Which oxide, when added to water, can produce a solution that is a strong acid?

(A) Na_2O

(B) CaO

(C) Fe_2O_3

(D) N_2O_5

(E) P_4O_{10}

58. Which oxide has the highest melting point?

(A) CO_2

(B) Al_2O_3

(C) Na_2O

(D) MgO

(E) SO_2

59. Which is the best description of the chemical change that occurs when magnesium metal dissolves in dilute hydrochloric acid?

(A) reduction of Mg

(B) oxidation of Mg

(C) proton donation by HCl

(D) proton acceptance by Cl^-

(E) proton donation by Cl^-

60. These isoelectronic species are listed in order of increasing atomic mass. When the list is rearranged according to increasing positive charge-to-mass ratio, what position is occupied by $^{39}K^+$?

$$^{31}P^{3-} \quad ^{35}Cl^- \quad ^{39}K^+ \quad ^{40}Ar^0 \quad ^{45}Sc^{3+}$$

(A) first

(B) second

(C) third

(D) fourth

(E) fifth

61. Which applies to the formation of solid water (ice) crystals on the surface of a package of frozen food stored in a home freezer?

I. Hydrogen bonds form between water molecules.

II. Energy is absorbed from the food package by the water molecules

III. The entropy of the system decreases

(A) I only

(B) III only

(C) I and III only

(D) II and III only

(E) I, II and III

62. The heat of solution for $LiBr_{(s)}$, a crystalline solid, is -48.8 kJ mol^{-1}. Which describes the changes that are predicted to occur when $LiBr_{(s)}$ dissolves in water?

(A) The temperature of the system remains the same as the entropy increases.

(B) The temperature of the system increases as the entropy increases.

(C) The temperature of the system decreases as the entropy increases.

(D) The temperature of the system increases as the entropy decreases.

(E) The temperature of the system decreases as the entropy decreases.

63. Which is the best description of the chemical change that occurs when excess $OH^-_{(aq)}$ is added to a solution of KHC_2O_4?

    (A) A proton is transferred from each $HC_2O_4^-$ ion to form H_2O molecules.

    (B) A proton is transferred from each $HC_2O_4^-$ ion to form H_3O^+ ions.

    (C) A proton is transferred from each $HC_2O_4^-$ ion to release CO_2 molecules.

    (D) A proton is transferred to each $HC_2O_4^-$ ion to form $H_2C_2O_4$ molecules.

    (E) Two protons are transferred to each $HC_2O_4^-$ ion to form $H_3C_2O_4^+$ ions.

64. The heat of solution for $NH_4NO_{3(s)}$ is +25.7 kJ mol^{-1}. Its solubility is 120 g per 100 g H_2O at 298 K. Which gives the best description of the thermodynamic parameters ΔH, ΔG and ΔS, for the dissolving process at 298 K?

	ΔH	ΔG	ΔS
(A)	>0	<0	>0
(B)	<0	<0	<0
(C)	>0	>0	>0
(D)	>0	0	>0
(E)	>0	0	<0

65. What is the best description of the distribution of bonding electrons at the carbon atom in a molecule of formic acid, HCOOH?

    (A) one *pi* bond and two *sigma* bonds with *sp* hybridization

    (B) one *pi* bond and three *sigma* bonds with sp^2 hybridization

    (C) one *pi* bond and three *sigma* bonds with sp^3 hybridization

    (D) four *sigma* bonds with sp^3 hybridization

    (E) three *sigma* bonds with sp^2 hybridization

66. All of these properties are the same for both isotopes of chlorine, $^{35}_{17}Cl$ and $^{37}_{17}Cl$, EXCEPT

    (A) nuclear mass

    (B) nuclear charge

    (C) extra-nuclear charge

    (D) number of occupied orbitals in the outer energy level

    (E) number of vacant orbitals in the outer energy level

67. When 100 mL of 0.100 M $Ba(NO_3)_2$ solution is added to 100 mL of 0.100 M NaF solution, a white precipitate forms. Which ratio of concentrations of ions in the resulting solution is closest to 2:1?

(A) $\dfrac{[F^-]}{[NO_3^-]}$

(B) $\dfrac{[F^-]}{[Ba^{2+}]}$

(C) $\dfrac{[NO_3^-]}{[Na^+]}$

(D) $\dfrac{[Ba^{2+}]}{[Na^+]}$

(E) $\dfrac{[NO_3^-]}{[Ba^{2+}]}$

68. Magnesium sulfate, $MgSO_4$ (molar mass: 120 g) occurs as a hydrated salt, $MgSO_4$ $7H_2O$ (molar mass: 246 g). Which expression gives the gain in mass of the solid phase expected when 10.0 grams of anhydrous magnesium sulfate is dissolved in water and the solution allowed to evaporate?

(A) $10.0 \times \dfrac{246}{120}$

(B) $10.0 \times \dfrac{126}{246}$

(C) $10.0 \times \dfrac{246}{126}$

(D) $10.0 \times \dfrac{126}{120}$

(E) $10.0 \times \dfrac{120}{126}$

69. Which describes the concentrations of K^+ and Cl^- ions in the reaction mixture produced when $KOH_{(aq)}$ is added to $HCl_{(aq)}$?

	[K^+]	[Cl^-]
(A)	decreases	decreases
(B)	remains the same	decreases
(C)	increases	remains the same
(D)	remains the same	decreases
(E)	increases	decreases

70. Which cation, when added to a solution of KOH, produces a colored precipitate?

    (A) Ca^{2+}

    (B) K^+

    (C) Mg^{2+}

    (D) Ni^{2+}

    (E) Zn^{2+}

71. In a dilute aqueous solution of H_3PO_4, which proton donor has the highest concentration?

    (A) H_3O^+

    (B) $HPO_4{}^{2-}$

    (C) H_3PO_4

    (D) $H_2PO_4{}^-$

    (E) $H_4PO_4{}^+$

72. When $BaSO_4$ dissolves to form a saturated solution, $[Ba^{2+}] = 1.05 \times 10^{-5}$. When the solubility product constant, K_{sp}, for $BaSO_4$ is expressed in scientific notation, what is the value of its exponential term?

    (A) -30

    (B) -25

    (C) -15

    (D) -10

    (E) -5

73. Two Lewis structures can be used to represent the resonance features of the bonding in the sulfur dioxide, SO_2, molecule. Which Lewis structure below is one of those resonance structures?

    (A) 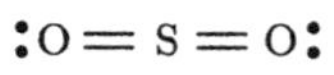 $:O=S=O:$

    (B) 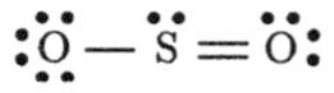 $:\ddot{O}-\ddot{S}=\ddot{O}:$

    (C) 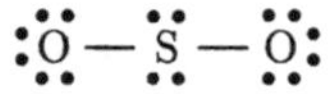 $:\ddot{O}-\ddot{S}-\ddot{O}:$

    (D) 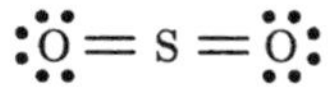 $:\ddot{O}=S=\ddot{O}:$

    (E) 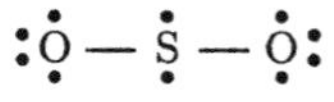 $:\dot{O}-\dot{S}-\dot{O}:$

74. What is the sum of all coefficients when the following equation is balanced using smallest possible integers?

$$...C_{12}H_{22}O_{11(s)} + ...O_{2(g)} \rightarrow ...CO_{2(g)} + ...H_2O_{(g)}$$

(A) 4

(B) 12

(C) 13

(D) 35

(E) 36

75. Spontaneous dissolving processes for ionic compounds in water can be exothermic or endothermic. Which component of the dissolving process supplies most of the energy that is produced in an exothermic process?

(A) dissociation of the solvent

(B) increase in entropy

(C) hydration of ions

(D) ionization energy of the cation

(E) loss of energy of activation

Section II

Section II - Free Response   Total Time – 90 Minutes
(Multiple-Choice Questions are found in Section I.)

Part A: Question 76
and
Question 77 or Question 78
Time: 40 minutes

Access to calculators, Periodic Table, lists of standard reduction potentials, and
Equations and Constants

(2004 Examination directions) Clearly show the method used and the steps involved in arriving at your answers. It is to your advantage to do this, because you may obtain partial credit if you do and you will receive little or no credit if you do not. Attention should be paid to significant figures. Be sure to write all your answers to the questions on the lined pages following each question in the booklet with the pink cover. Do not write your answers on the green insert.

Answer question 76 below. The Section II score weighting for this question is 20 percent.

76. Ammonium hydrogen sulfide is a white crystalline solid that decomposes according to the equation

$$NH_4HS_{(s)} \rightleftharpoons NH_{3(g)} + H_2S_{(g)}$$

In one experiment, solid NH_4HS was placed in a 5.0 L rigid container at 298 K. At equilibrium, the total pressure was 0.660 atm with the gas phase in contact with excess solid.

(a) Write the mass action expression, K_p, for this equilibrium system.

(b) Calculate the numerical value for K_p at 298 K.

(c) Equilibrium in this system was disturbed by the addition of $NH_{3(g)}$. A new equilibrium including the solid phase was established in which the pressure of $NH_{3(g)}$ was equal to three times the pressure of $H_2S_{(g)}$: $P_{NH3} = 3\,P_{H2S}$

    (i) Calculate the partial pressure of H_2S at this new equilibrium.

    (ii) Calculate the change in moles of $NH_4HS_{(s)}$ present in the system; include the correct sign for this change.

(d) In a different experiment in a different vessel at the same temperature, $NH_{3(g)}$ and $H_2S_{(g)}$ were mixed At the time of mixing, the partial pressure of H_2S was 1.00 atm. The initial partial pressure of $NH_{3(g)}$ was unknown. After equilibrium was established, the partial pressure of H_2S was 0.75 atm.

    (i) Calculate the partial pressure of NH_3 at equilibrium.

    (ii) Calculate the original pressure of NH_3 at the time of mixing.

Answer either question 77 or question 78 below.

(2004 examination directions) Only one of these two questions will be graded. If you start both questions, be sure to cross out the question you do not want graded.

The Section II score weighting for the question that you choose is 20 percent.

77. The following data were obtained from a study of the kinetics of the reaction below at 298 K.

$$5Br^- + BrO_3^- + 6H^+ \rightarrow 3Br_2 + 3H_2O$$

	initial concentration mol L^{-1}			rate of formation moles sec^{-1}
Trial	**[Br$^-$]**	**[BrO$_3$$^-$]**	**[H$^+$]**	**Br$_2$**
I.	1.0×10^{-3}	5.0×10^{-3}	$10. \times 10^{-3}$	2.5×10^{-4}
II.	2.0	5.0	10.	5.0
III.	1.0	10.	10.	2.5
IV.	1.0	5.0	20.	10.
V.	2.0	10.	20.	?

(a) Write the rate law for this reaction. What is the overall order for this reaction?

(b) Calculate the rate constant, k, for this reaction. Specify units.

(c) What is the predicted initial rate of formation of bromine in trial V?

(d) When trial III has reached completion, the concentration of one of the dissolved species is greater than either of the other two dissolved species.

  (i) Which of the dissolved species has the highest concentration? Explain your choice.

  (ii) Calculate that concentration.

78. The following problems concern quantitative analysis of two chemical compounds.

   (a) A compound known to contain only the elements carbon, hydrogen, nitrogen, and oxygen was analyzed in the laboratory.

       (i) A sample of the compound with mass 0.4788 g was sent through a series of tests that converted all combined nitrogen into nitrogen gas. The nitrogen gas was collected by water displacement and yielded a volume of 37.80 mL, measured at 23.8°C and 746.0 mmHg. According to a chemical handbook, at this temperature, the vapor pressure of water is 22.1 mmHg. Using the results of this experiment, calculate the mass percent of nitrogen in the compound.

       (ii) In a separate experiment, 12.96 mg of the compound was burned in a pure oxygen atmosphere. Products collected were 35.14 mg carbon dioxide and 8.638 mg water. Using the results of this experiment, calculate the mass percent of carbon and hydrogen in the compound.

       (iii) Explain how to use data from both experiments to calculate the mass percent of oxygen in the compound.

   (b) In a separate experiment, a different compound is shown to consist of 25.4% by mass carbon, 3.20% by mass hydrogen, 37.5% by mass chlorine, and 33.9% by mass oxygen.

       (i) Determine the empirical formula of the compound.

       (ii) Identify additional information about the compound that is needed in order to determine the molecular formula.

Part B: Questions 79, 80, 81 and
Question 82 or Question 83
Time: 50 minutes

Access to Periodic Table, lists of standard reduction potentials
and *Equations and Constants*
No access to calculators

Answer question 79 below: The Section II score weighting for this question is 15 percent.

79. (2004 Examination directions) Write the formulas to show the reactants and the products for any FIVE of the laboratory situations described below. In all cases a reaction occurs. Assume that solutions are aqueous unless otherwise indicated. Represent substances in solution as ions if the substances are extensively ionized. Omit formulas for any ions or molecules that are unchanged by the reaction. You need not balance the equations.

(a) Excess concentrated sodium hydroxide is poured onto solid aluminum hydroxide.

(b) Acidified solutions of iron(II) sulfate and potassium permanganate are mixed.

(c) A sample of 2-decanol is burned in excess oxygen.

(d) Solutions of dilute sulfuric acid and strontium hydroxide are mixed.

(e) Solutions of silver nitrate and sodium dichromate are mixed.

(f) A piece of calcium is heated in an atmosphere of pure nitrogen.

(g) A small quantity of dilute potassium hydroxide solution is poured into dilute nitrous acid solution.

(h) Solutions of tin(II) nitrate and cobalt(III) nitrate are mixed.

(2004 Examination directions) Your responses to the rest of the questions in this part of the examination will be graded on the basis of the accuracy and relevance of the information cited. Explanations should be clear and well organized. Examples and equations may be included in your responses where appropriate. Specific answers are preferable to broad, diffuse responses.

(2004 examination directions) **Answer both Question 80 and Question 81 below.** Both questions will be graded.

The Section II score weighting for these questions is 30 percent (15 percent each).

80. In order to determine the degree of hydration of Epsom salts, a hydrated form of magnesium sulfate, a student performs several laboratory tests. The student weighs a clean, dry crucible and lid. The student then adds several grams of Epsom salts to the crucible and reweighs it. The crucible and salt sample are heated over a Bunsen burner flame for five minutes; the crucible is allowed to cool, then reweighed. The crucible is then returned to the Bunsen burner flame for an additional five minutes of heating. It is allowed to cool, then is reweighed.

    (a) Prepare a data table to specify the measurements that should be taken during this procedure.

    (b) How can the student tell when this lab procedure is completed? Explain.

    (c) If the crucible is initially overheated so that some of the salt appears as a white wisp drifting away from the crucible, what effect will this have on the data recorded and on the final determination of degree of hydration? Explain.

    (d) If the empty crucible selected initially is clean but not dry, what effect will this have on the final determination of degree of hydration? Explain.

81. Consider an aqueous solution that is 10% NaCl by mass.

    (a) Identify the additional facts about the solution that are needed (can be calculated or must be determined) in order to:

    (i) calculate the molality, m, of the NaCl in the solution.

    (ii) calculate the molarity, M, of the solution.

    (iii) calculate the mole fraction, χ, of water in the solution.

    (iv) determine the freezing point of the solution at 1 atm. Assume complete dissociation of the solute.

    (b) The original solution of NaCl is heated from 20°C to 30°C at constant pressure. Which of the four measures of concentration is affected? Explain.

    - molarity
    - molality
    - mass percent
    - mole fraction

(2004 examination directions) **Answer either question 82 or question 83 below.** Only one of these two questions will be graded. If you start both questions, be sure to cross out the question you do not want graded. The Section II score weighting for the question that you choose is 15 percent.

82.   Carbon monoxide and nitrogen monoxide are both gases that are found in small quantities in Earth's atmosphere.

    (a)   Which gas has molecules with the greater root-mean-square speed at 25°C? Explain.

    (b)   What is the ratio of the rates of effusion of CO to NO? Show a set-up for your calculation. A calculated numerical answer is not required.

    (c)   Draw a reasonable Lewis structure for each molecule. Which structure does not illustrate the usual principles for construction of Lewis structures? Explain.

    (d)   Both of these gases causes some pollution of the air due to burning fossil fuels in the internal combustion engines of automobiles? Explain.

83.   Although the elements scandium and selenium are both found in the Period 4 (fourth row) of the Periodic Table, the physical and chemical characteristics of these two elements are quite different. Explain the following differences in properties of the elements scandium and selenium in terms of atomic structure and/or bonding.

    (a)   The atomic radius of Sc is 160 pm while that of Se is 121 pm.

    (b)   Scandium forms only one oxidation state but selenium has several, ranging from -2 to $+6$.

    (c)   Scandium is a good conductor of electricity. Selenium is a poor conductor of electricity.

    (d)   The reaction product of scandium with pure oxygen is Sc_2O_3 while selenium reacts with pure oxygen to form SeO_2.

NOTES

NOTES

NOTES

NOTES

NOTES